Human Physiology
From Cells to Systems

NINTH EDITION

Lauralee Sherwood

West Virgina University

CENGAGE
Learning·

Australia · Brazil · Mexico · Singapore · United Kingdom · United States

ISBN: 978-1-305-27346-7

Cengage Learning
20 Channel Center Street
Boston, MA 02210
USA

Cengage Learning is a leading provider of customized learning solutions with office locations around the globe, including Singapore, the United Kingdom, Australia, Mexico, Brazil, and Japan. Locate your local office at: **www.cengage.com/global**.

Cengage Learning products are represented in Canada by Nelson Education, Ltd.

To learn more about Cengage Learning Solutions, visit **www.cengage.com**.

Purchase any of our products at your local college store or at our preferred online store **www.cengagebrain.com**.

Printed in the United States of America
1 2 3 4 5   19 18 17 16 15

Table of Contents

Preface

This coloring book has been specially designed to work with Lauralee Sherwood's Human Physiology textbook. Pieces of art have been selected from the text book to help improve your understanding of physiology concepts. Figure numbers have been included on each page to allow you to quickly reference your textbook for more help.

Using this book:

The figure captions and descriptions have been included to help you understand the processes depicted on each page. Reading these first will help you fill with the rest of the activities on the page. There are three activities in this coloring book: Coloring, labeling, and describing.

Start by coloring in each letter of a term, and then filling in the appropriate part of the figure. This will help you to associate parts of the figure with their term and also help you learn the proper spelling. Next, color in each labeling term, and then label that term on the figure. You might need to simply draw a line to indicate the correct part, or you might need to circle an area. Lastly, where appropriate, areas have been left for you to write down the steps associated with a process. Think about rewording the descriptions of the steps to make sure that you understand the process, and are not just copying it.

We hope you find this product useful as you explore the science of how our bodies work.

1-3 The stomach as an organ made up of all four primary tissue types.

Epithelial tissue
Connective tissue
Muscle tissue
Nervous tissue
Organ

What is the purpose of each type of tissue?

Epithelial tissue: _____

_____

Connective tissue: _____

_____

Muscle tissue: _____

_____

Nervous tissue: _____

_____

_____

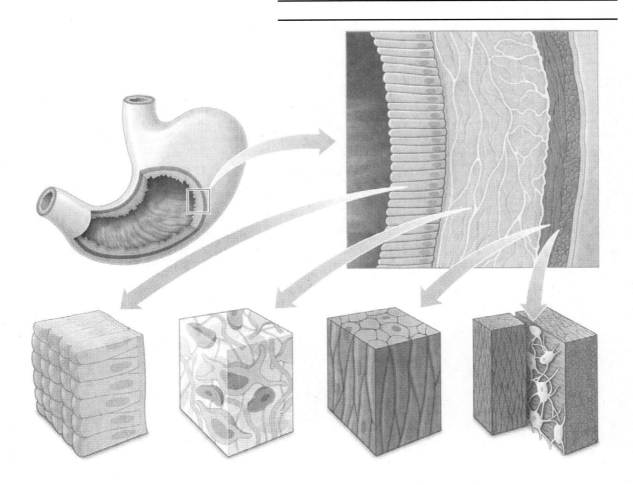

1-4 Exocrine and endocrine gland formation during development.

(a) Glands arise from the formation of pocket like invaginations of surface epithelial cells. (b) If the cells at the deepest part of the invagination become secretory and release their product through the connecting duct to the surface, an exocrine gland is formed. (c) If the connecting cells are lost and the deepest secretory cells release their product into the blood, an endocrine gland is formed.

Color

Surface epithelium

Surrounding epithelial
  cells

Duct Cells

Secretory exocrine
  gland cells

Connecting cells lost
  during development

Secretory endocrine
  gland cell

Blood vessel

Directional arrows

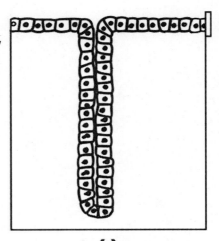

(a)

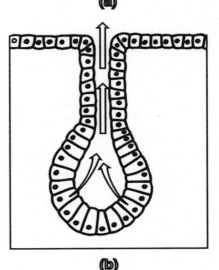

(b)

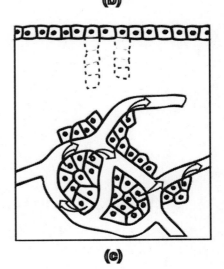

(c)

3

1-5 Components of the body systems.

(*Source:* Adapted from Cecie Starr and Ralph Taggart, Biology: The Unity and Diversity of Life, Eighth Edition, Fig. 33.11, pp. 552–553. Copyright © 1998 Wadsworth Publishing Company.)

Circulatory system
Digestive system
Respiratory system
Urinary system
Skeletal system
Muscular system

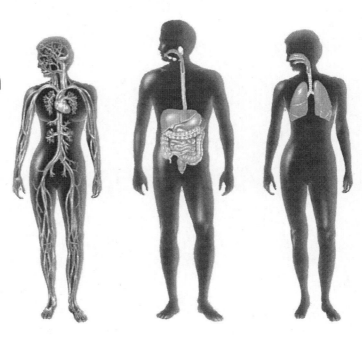

What are the components of each system?

Circulatory system
_____
_____
_____

Digestive system
_____
_____
_____

Respiratory system
_____
_____
_____

Urinary system
_____
_____
_____

Skeletal system
_____
_____
_____

Muscular system
_____
_____
_____

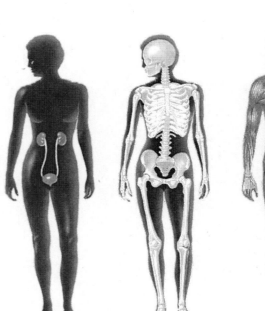

5

1-5 Components of the body systems (cont.)

(Source: Adapted from Cecie Starr and Ralph Taggart, Biology: The Unity and Diversity of Life, Eighth Edition, Fig. 33.11, pp. 552–553. Copyright © 1998 Wadsworth Publishing Company.)

Integumentary system
Immune system
Nervous system
Endocrine system
Reproductive system

What are the components of each system?

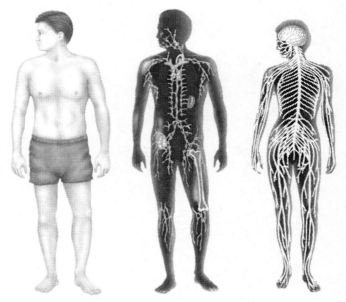

Integumentary system

_____
_____
_____

Immune system

_____
_____
_____

Nervous system

_____
_____
_____

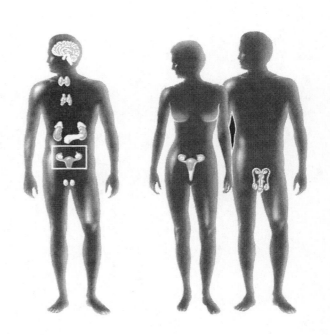

Endocrine system

_____
_____
_____

Reproductive system- Male

_____
_____
_____

Reproductive system- Female

_____
_____
_____

1-6 Components of extracellular fluid (internal environment).

Color
Cell
Interstitial fluid
Blood vessel
Plasma
Extracellular Fluid

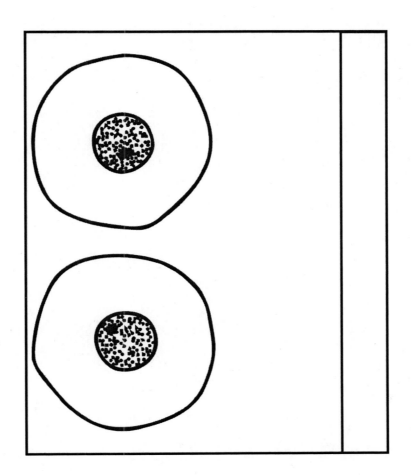

1-7 Interdependent relationship of cells, body systems, and homeostasis.

The depicted interdependent relationship serves as the foundation for modern-day physiology: *Homeostasis is essential for the survival of cells, body systems maintain homeostasis, and cells make up body systems.*

Color

Body Systems

Homeostasis

Cells

Directional arrows

Label

Body systems

Homeostasis

Directional arrows

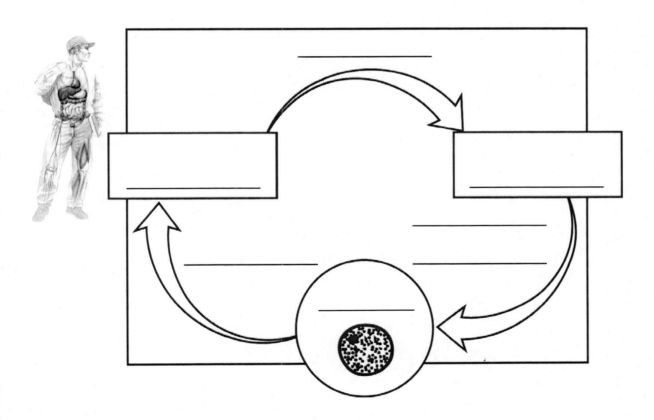

2-1 Diagram of cell structures visible under an electron microscope.

Color

Centriole

Endocytotic vesicle

Exocytotic vesicle

Free Ribosome

Gesicle

Golgi complex

Lysosome

Mitochondrion

Nuclear Pores

Nucleus

Peroxisome

Plasma
    membrane

Rough
    endoplasmic
    reticulum

Smooth
    endoplasmic
    reticulum

Label

Cytosol

Microfilaments

Microtubule

Vault

Ribosomes

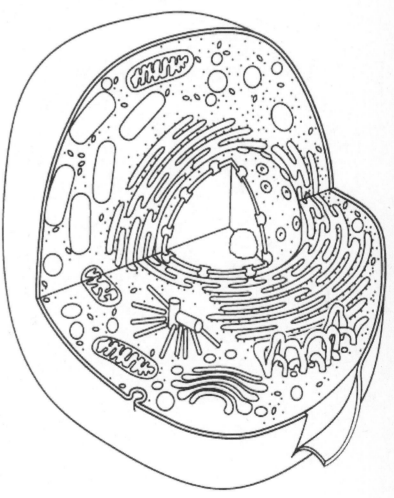

2-3 Overview of the secretion process for proteins synthesized by the endoplasmic reticulum.

Note that the secretory product never comes into contact with the cytosol.

Golgi complex
Lysosome
Nucleus
Proteins
Ribosomes
Rough ER
Secretion
    (exocytosis)
Secretory
    vesicles
Smooth ER
Transport
    vesicles

List the 7 steps in this process

1._____
_____
_____

2._____
_____
_____

3._____
_____
_____

4._____
_____

5._____
_____

6._____
_____

7._____
_____

2-5 Golgi complex.

Diagram and electron micrograph of a Golgi complex, which consists of a stack of slightly curved, membrane-enclosed sacs. The vesicles at the dilated edges of the sacs contain finished protein products packaged for distribution to their final destination.

Color
Transport vesicle

Vesicles containing finished product

Golgi lumen

Ribosomes

Label
Golgi complex

Golgi sacs

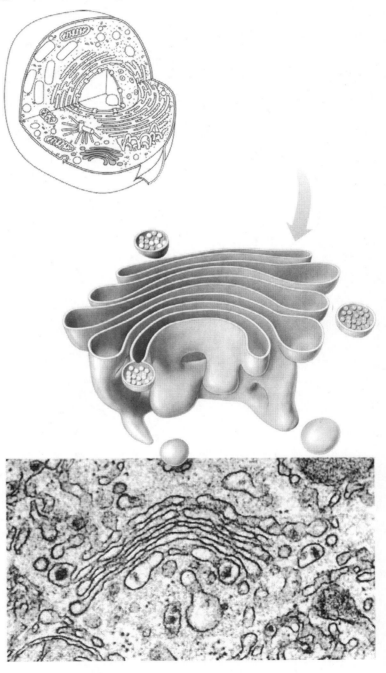

2-6 Exocytosis and endocytosis.

(a) Exocytosis: A secretory vesicle fuses with the plasma membrane, releasing the vesicle contents to the cell exterior. The vesicle membrane becomes part of the plasma membrane. (b) Endocytosis: Materials from the cell exterior are enclosed in a segment of the plasma membrane that pockets inward and pinches off as an endocytic vesicle.

Color

Cytosol

Extracellular fluid (ECF)

Vesicle contents

Plasma membrane

Label

Endocytic Vesicle

Secretory vesicle

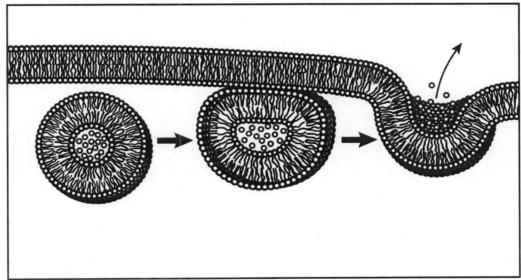

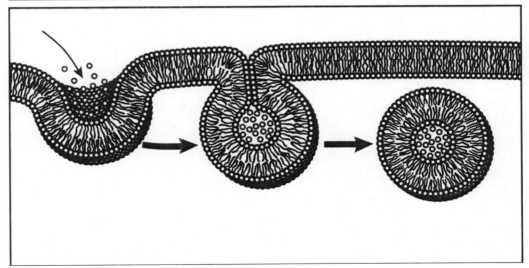

2-7 Packaging, docking, and release of secretory vesicles.

The diagram series illustrates secretory vesicle formation and budding with the aid of a coat protein and docking with the plasma membrane by means of v-SNAREs and t-SNARES. The transmission electron micrograph shows secretion by exocytosis.

Color

Extracellular fluid (ECF)

Plasma Membrane

Cytosol

Golgi Lumen

Membrane of outermost
        Golgi sac

Recognition markers

t-SNARE

Coat-protein acceptor

v-SNARE

Sorting signal

Cargo Proteins

List the 4 steps in this process

1._____

_____

2._____

_____

3._____

_____

4._____

_____

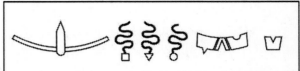

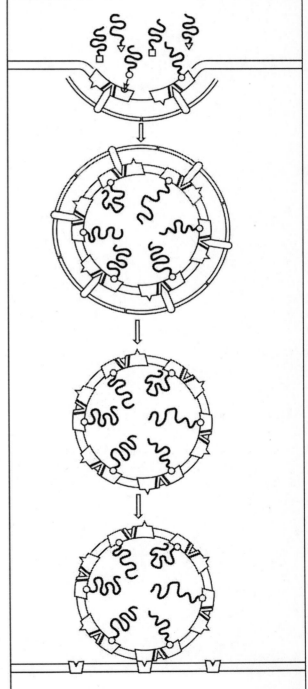

2-8 Lysosomes and peroxisomes.

Diagram and electron micrograph of lysosomes, which contain hydrolytic enzymes, and peroxisomes, which contain oxidative enzymes.

Color

Peroxisome

Lysosome

Oxidative
    enzymes

Hydrolytic
    enzymes

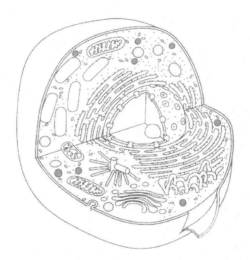

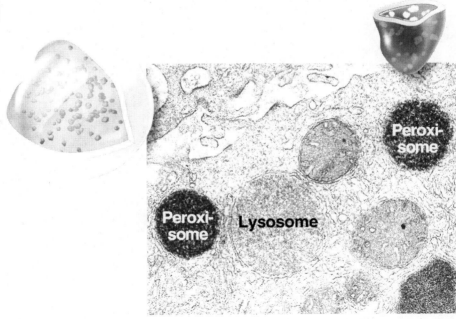

Peroxi-
some

Peroxi-
some

Lysosome

2-9 Forms of endocytosis.

(a) Diagram and electron micrograph of pinocytosis. The surface membrane dips inward to form a pouch, then seals the surface, forming an intracellular endocytic vesicle that nonselectively internalizes a bit of ECF. (b) Diagram and electron micrograph of receptor-mediated endocytosis. When a large molecule such as a protein attaches to a specific surface receptor, the membrane pockets inward with the aid of a coat protein, forming a coated pit, then pinches off to selectively internalize the molecule in an endocytic vesicle. (c) Diagram and scanning electron micrograph series of phagocytosis. White blood cells internalize multimolecular particles such as bacteria or old red blood cells by extending pseudopods that wrap around and seal in the targeted material. A lysosome fuses with and degrades the vesicle contents.

Color

Protein

Nucleus of cell

Surface receptor site

Endocytotic pouch

Endocytotic vesicle

Cytosol

White Blood Cell

Pseudopod

Bacterium

Phagocytic vesicle

Lysosome

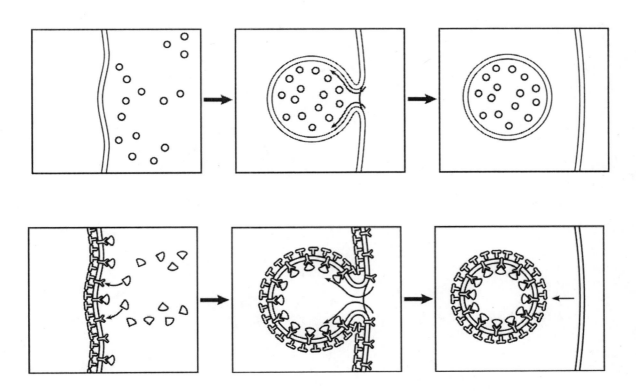

2-10 Mitochondrion.

Diagram and electron micrograph of a mitochondrion. Note that the outer membrane is smooth while the inner membrane forms folds known as cristae that extend into the matrix. An intermembrane space separates the outer and inner membranes. The electron transport proteins embedded in the cristae are ultimately responsible for converting much of the energy of food into a usable form.

Color

Mitochondrion

Intermembrane space

Inner mitochondrial
    membrane

Outer mitochondrial
    membrane

Proteins of electron
    transport system

Matrix

Cristae

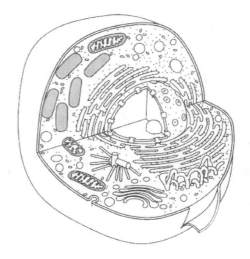

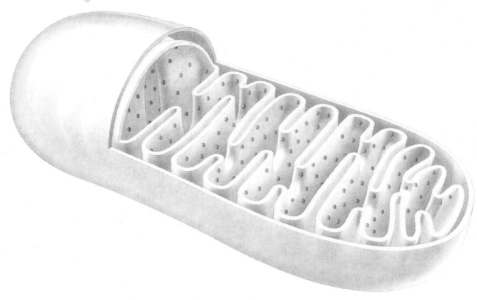

2-11 Stages of cellular respiration.

The three stages of cellular respiration are (1) glycolysis, (2) citric acid cycle, and (3) oxidative phosphorylation.

Color and Label

Citric acid cycle

Oxidative phosphorylation

Mitochondrial matrix

Mitochondrial inner membrane

Glycolysis

Glucose and other fuel molecules

Cytosol

Pyruvate

2 ATP

28 ATP

Acetyl-CoA

Pyruvate to acetate

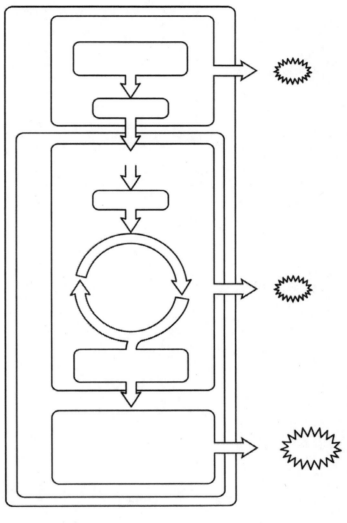

Explain the 3 stages involved in cellular respiration.

1._____

_____

_____

2._____

_____

_____

3._____

_____

_____

2-13 Citric acid cycle in the mitochondrial matrix.

The two carbons entering the cycle by means of acetyl-CoA are eventually converted to CO2, with oxaloacetate, which accepts acetyl-CoA, being regenerated at the end of the cyclical pathway. The hydrogens released at specific points along the pathway bind to the hydrogen carrier molecules NAD1 and FAD for further processing by the electron transport system. One molecule of ATP is generated for each molecule of acetyl-CoA that enters the citric acid cycle, for a total of two molecules of ATP for each molecule of processed glucose.

Color

Cytosol

Mitochonrdial
        membranes

Mitochodrial
        matrix

Crista

Color and Label

Pyruvic acid

Acetic acid

Acytely CoA

Oxaloacetic acid

Citric Acid

Isocitric acid

α Ketoglutaric
        acid

Succinyl CoA

Succinic acid

Fumaric acid

Malic acid

ATP

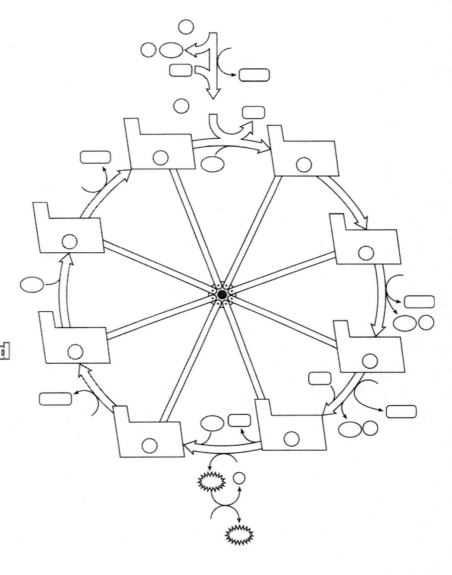

2-14 Oxidative phosphorylation at the mitochondrial inner membrane.

(a) ATP synthesis resulting from the passage of high-energy electrons through the mitochondrial electron transport chain.

Color

Mitochondrial matrix

Cytosol

Outer mitochondrial
    membrane

Headpiece

Basal Unit

Stalk

Intermembrane space

Inner mitochondrial
    membrane

ATP

Hydrogen ions (H⁺)

Electrons (e⁻)

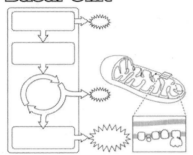

Label

Path of electron (e⁻)
    transport

Path of hydrogen ions (H⁺)

ADP = P_i

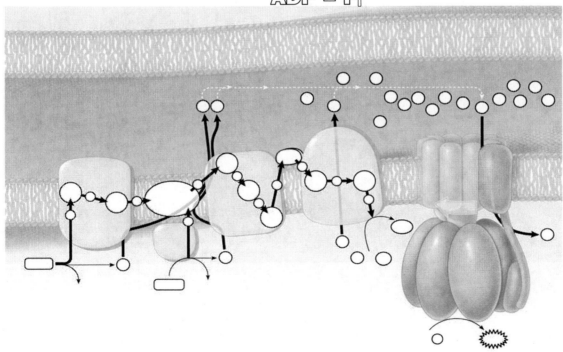

2-14 Oxidative phosphorylation at the mitochondrial inner membrane (cont).

Explain the 9 steps involved in this process.

1. _____
   _____
   _____
   _____

2. _____
   _____
   _____
   _____

3. _____
   _____
   _____
   _____

4. _____
   _____
   _____
   _____

5. _____
   _____
   _____
   _____

6. _____
   _____
   _____
   _____

7. _____
   _____
   _____
   _____

8. _____
   _____
   _____
   _____

9. _____
   _____
   _____

2-17 Vaults.

Diagram of closed and open vaults and electron micrograph of vaults, which are octagonal barrel-shaped nonmembranous organelles believed to transport either messenger RNA or the ribosomal subunits from the nucleus to cytoplasmic ribosomes.

Color

Closed vault

Open Vault

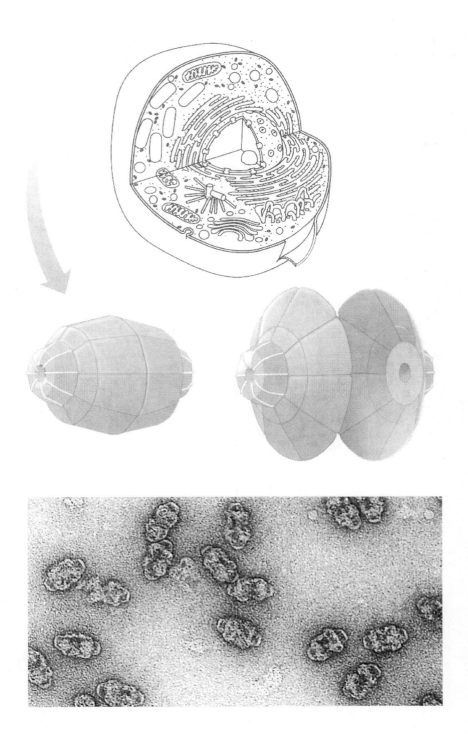

2-20 Centrioles.

The two cylindrical centrioles of the pair lie at right angles to each other as shown in the diagram. The electron micrograph shows a centriole in cross section. Note that a centriole is made up of nine microtubule triplets that form a ring.

Color

Microtubule triplet

2-21 Two-way vesicular axonal transport facilitated by the microtubular "highway" in a nerve cell.

Secretory vesicles are transported from the site of production in the cell body along a microtubule "highway" to the terminal end for secretion. Vesicles containing debris are transported in the opposite direction for degradation in the cell body. The top enlargement depicts kinesin, a molecular motor, carrying a secretory vesicle down the microtubule by using its "feet" to "step" on one tubulin molecule after another. The bottom enlargement depicts another molecular motor, dynein, transporting the debris up the microtubule.

Color

Cell body
Endoplasmic reticulum
Nucleus
Golgi complex
Microtubular "highway"
Secretory vesicle
lysosome
Axon
Debris
Axon Terminal
Secretory vesicle
Kinesin molecule
Microtubule

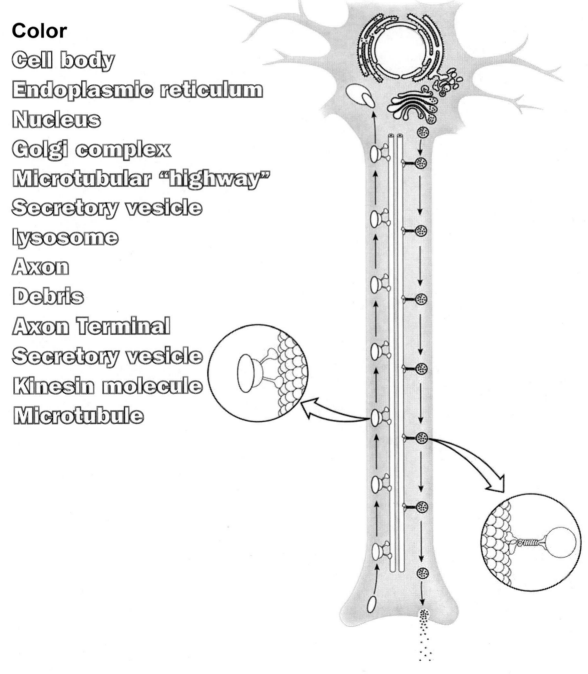

2-23 Structure of a cilium or flagellum.

(a) The relationship between the microtubules and the centriole turned basal body of a cilium or flagellum. (b) Diagram of a cilium or flagellum in cross section showing the characteristic "9 1 2" arrangement of microtubules along with the dynein arms and other accessory proteins that hold the system together.

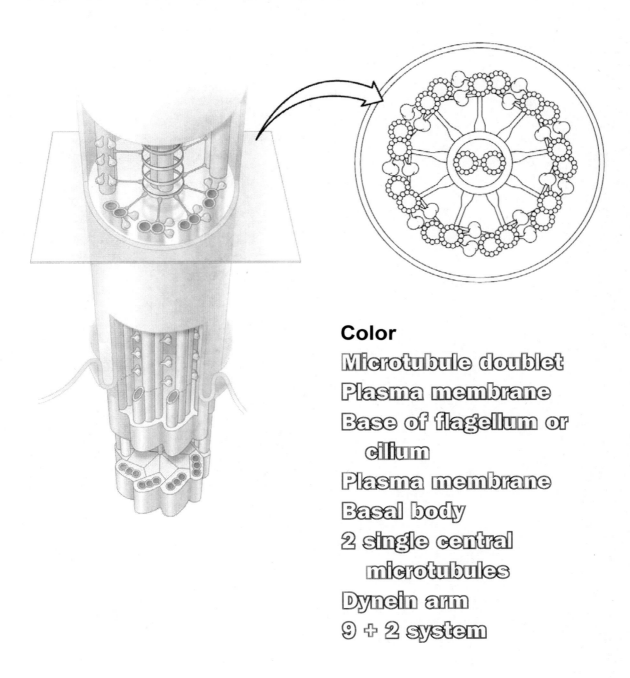

Color
Microtubule doublet

Plasma membrane

Base of flagellum or
   cilium

Plasma membrane

Basal body

2 single central
   microtubules

Dynein arm

9 + 2 system

2-24 Cytokinesis.

(a) Schematic illustration of the actin contractile ring squeezing apart the two duplicate cell halves during cytokinesis.

Color

Extracellular fluid (ECF)

Intracellular Fluid (ICF)

Nucleus

Contractile ring
composed of actin

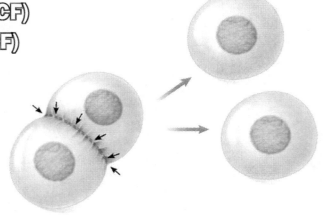

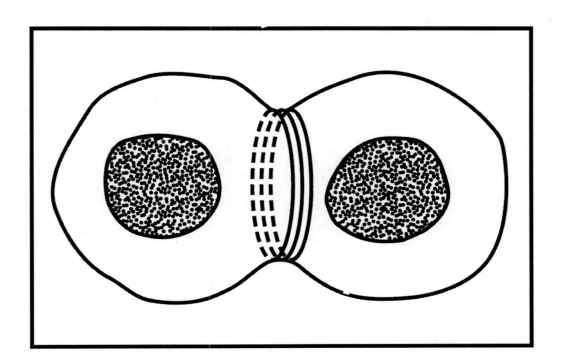

3-2 Structure and organization of phospholipid molecules in a lipid bilayer.

(a) Phospholipid molecule. (b) In water, phospholipid molecules organize themselves into a lipid bilayer with the polar heads interacting with the polar water molecules at each surface and the nonpolar tails all facing the interior of the bilayer. (c) An exaggerated view of the plasma membrane enclosing a cell, separating the ICF from the ECF. (Source: Adapted from Cecie Starr and Ralph Taggart, Biology: The Unity and Diversity of Life, Eighth Edition, Fig. 4-2c, p. 56. Copyright © 1998 Wadsworth Publishing Company.)

Color

Choline

Phosphate

Glycerol

Fatty acid

Tails (nonpolar, hydrophobic)

Head (polar, hydrophilic)

Intracellular fluid

Extracellular fluid

Lipid bilayer

Negative charge on phosphate group

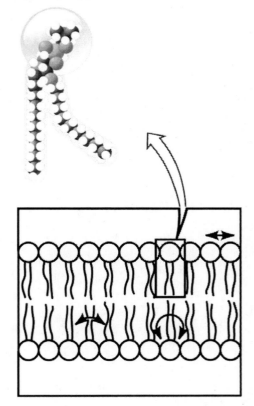

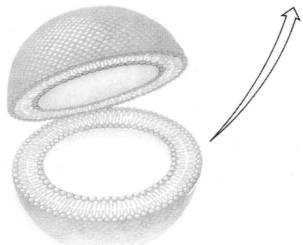

3-3 Fluid mosaic model of plasma membrane structure.

The plasma membrane is composed of a lipid bilayer embedded with proteins. Some of these proteins extend through the thickness of the membrane, some are partially submerged in the membrane, and others are loosely attached to the surface of the membrane. Short carbohydrate chains are attached to proteins or lipids on the outer surface only.

Color

Integral proteins

Extracellular fluid

Intracellular fluid

Carbohydrate chain

Phospholipid molecule

Cholesterol molecule

Peripheral proteins

Glycoprotein

Gated channel protein

Receptor protein

Leak channel protein

Carrier protein

Glycolipid

Cell adhesion molecule

Microfilament of cytoskeleton

Label

Lipid bilayer

Dark line

Light space

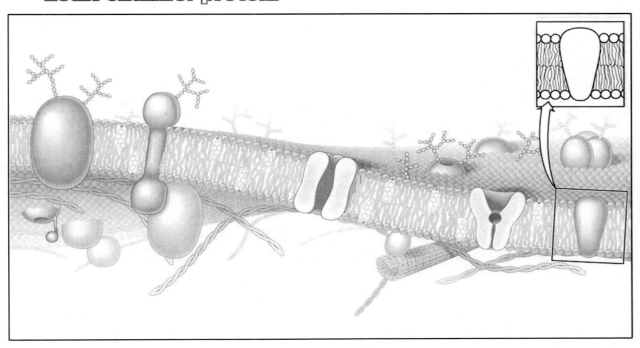

3-7 Diffusion.

(a) Diffusion down a concentration gradient. (b) Dynamic equilibrium, with no net diffusion occurring.

3-8 Diffusion through a membrane.

(a) Net diffusion across the membrane down a concentration gradient. (b) No diffusion through the membrane despite the presence of a concentration gradient.

Color

Solute molecule

Solvent

Membrane

If a substance can permeate the membrane:

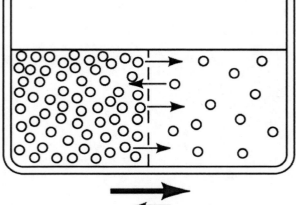

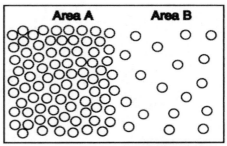

Area A   Area B

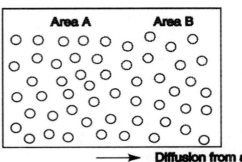

⟶   **Diffusion from area A to area B**

←   Diffusion from area B to area A

⟶   Net diffusion

If the membrane is impermeable to a substance:

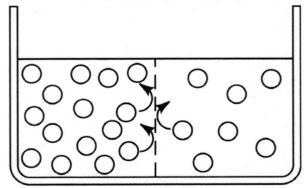

Area A   Area B

⟶   **Diffusion from area A to area B**

◀—   Diffusion from area B to area A

No net diffusion

O = Solute molecule   Net diffusion = diffusion from area A to area B minus diffusion from area B to area A

3-10 Osmosis when pure water is separated from a solution containing a non penetrating solute.

3-11 Movement of water and a penetrating solute unequally distributed across a membrane.

Color

Water molecule

Solute molecule

Explain what is happening below.

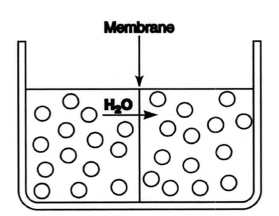

Membrane

H_2O

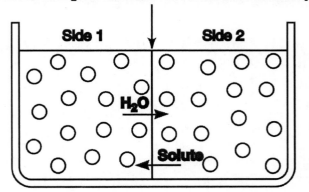

Membrane (permeable to both water and solute)

Side 1   Side 2

H_2O

Solute

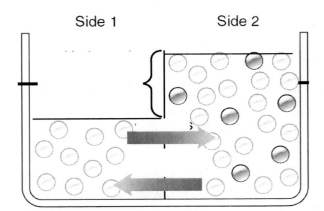

Side 1   Side 2

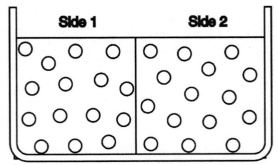

Side 1   Side 2

3-13 Tonicity and osmotic water movement.

Color

Water

Solute

Cells

Label

Isotonic conditions

Hypotonic conditions

Hypertonic conditions

Explain what happens in each solution:

(a) Isotonic conditions

_____

_____

_____

(b) Hypotonic conditions

_____

_____

_____

(c) Hypertonic conditions

_____

_____

_____

3-14 Facilitated diffusion, a passive form of carrier-mediated transport.

Color

Molecule to be transported

Carrier molecule

Phospholipid membrane

Direction of transport arrows

List the 4 steps in this process

1. _____

_____

_____

_____

2. _____

_____

_____

_____

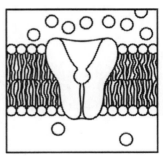

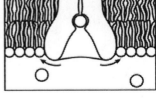
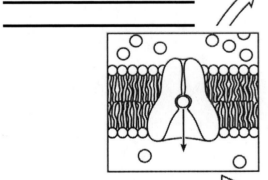

3. _____

_____

_____

_____

4. _____

_____

_____

_____

© 2016 Cengage Learning. All Rights Reserved. May not be scanned, copied or duplicated, or posted to a publicly accessible website, in whole or in part.

3-16 Na+–K+ pump.

The plasma membrane of all cells contains an active-transport carrier, the Na+–K+ pump, which uses energy in the carrier's phosphorylation–dephosphorylation cycle to sequentially transport Na+ out of the cell and K1 into the cell against these ions' concentration gradient. This pump moves three Na1 out and two K+ in for each ATP split.

Color

Molecule to be transported

Extracellular fluid (ECF)

Intracellular Fluid (ICF)

Carrier molecule

Phospholipid membrane

List the 6 steps in this process:

1._____

_____

_____

2._____

_____

_____

3._____

_____

_____

4._____

_____

_____

5._____

_____

_____

6._____

_____

_____

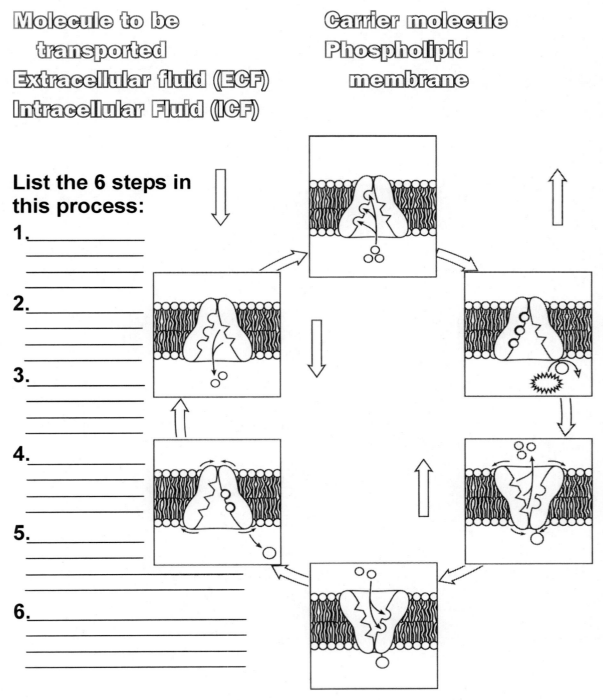

3-17 Secondary active transport, in which an ion concentration gradient is used as the energy source for active transport of a solute.

(a) In symport, the transported solute moves in the same direction as the gradient of the driving ion. (b) In antiport, the transported solute moves in the direction opposite from the gradient of the driving ion.

Color

Driving ion

Transported solute

Membrane

Carrier protein

Extracellular fluid (ECF)

Intracellular Fluid (ICF)

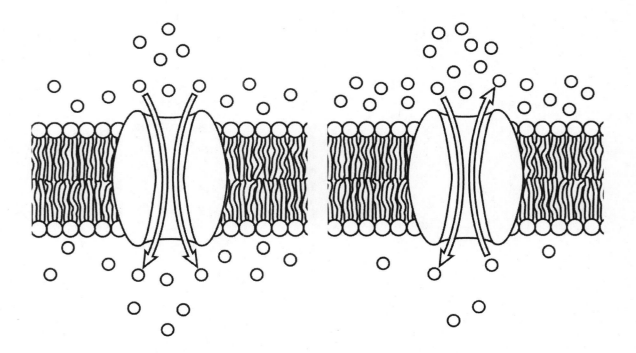

3-18 Symport of glucose.

Glucose is transported across intestinal and kidney cells against its concentration gradient by means of secondary active transport mediated by the sodium and glucose cotransporter (SGLT) at the cells' luminal membrane.

Color

Blood vessel

Lumen

Sodium- Potassium
   pump

GLUT

SGLT

Extracellular fluid (ECF)

Intracellular Fluid (ICF)

Glucose molecule

Sodium molecule

Label and describe the stages of this process:

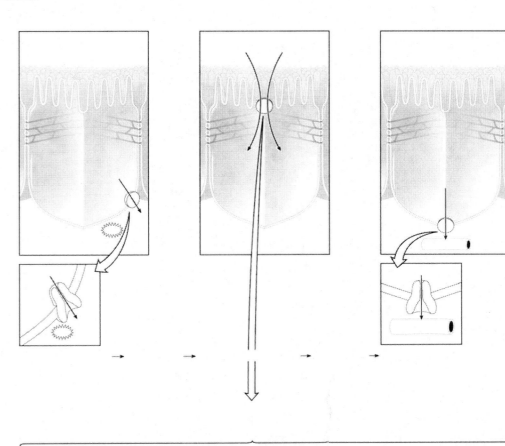

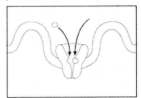
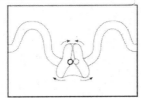
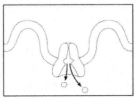

3-20 Equilibrium potential for K+.

Color

Plasma membrane

Extracellular fluid (ECF)

Intracellular Fluid (ICF)

Potassium

Anion

Label

Concentration gradient
for potassium

Electrical gradient for
potassium

List the 5 steps in this process:

1._____
_____

2._____
_____

3._____
_____

4._____
_____

5._____
_____

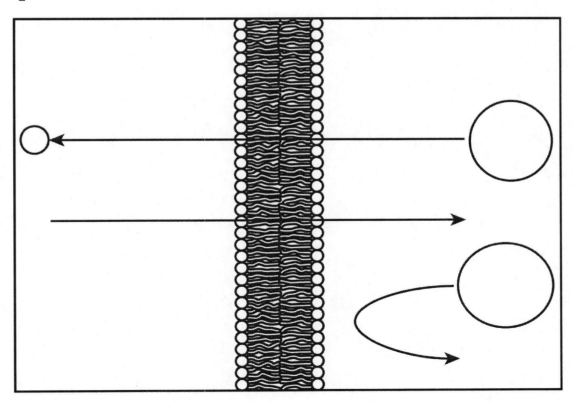

3-21 Equilibrium potential for Na+.

Color

Plasma membrane

Extracellular fluid (ECF)

Intracellular Fluid (ICF)

Sodium

Chloride

Label

Concentration gradient
for sodium

Electrical gradient for
sodium

List the 5 steps in this process:

1._____
_____

2._____
_____

3._____
_____

4._____
_____

5._____
_____

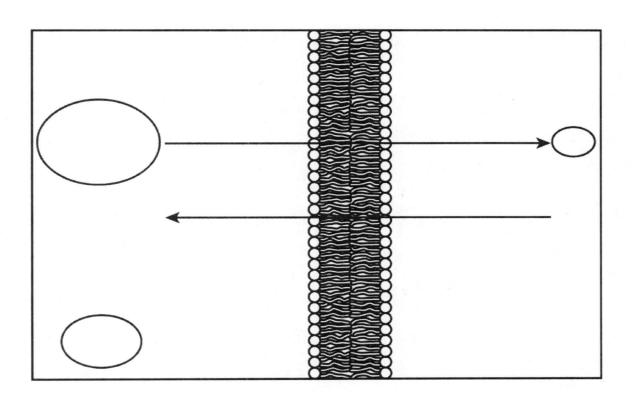

3-22 Effect of concurrent K+ and Na+ movement on establishing the resting membrane potential.

Color

Plasma membrane

Extracellular fluid (ECF)

Intracellular Fluid (ICF)

Sodium

Chloride

Potassium

Anion

List the 5 steps in this process:

1._____

_____

2._____

_____

3._____

_____

4._____

_____

5._____

_____

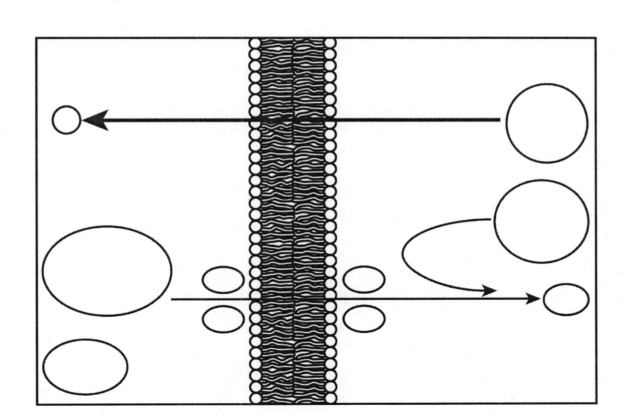

4-2 Current flow during a graded potential.

(a) The membrane of an excitable cell at resting potential. (b) A triggering event opens Na+ channels, leading to the Na+ entry that brings about depolarization. The adjacent inactive areas are still at resting potential. (c) Local current flow occurs between the active and adjacent inactive areas. This local current flow results in depolarization of the previously inactive areas. In this way, the depolarization spreads away from its point of origin.

Color

Extracellular fluid (ECF)

Intracellular fluid (ICF)

Na+ Channel

Excitable cell

Triggering event

Inactive area

Active area

Area being depolarized

(+) Protons

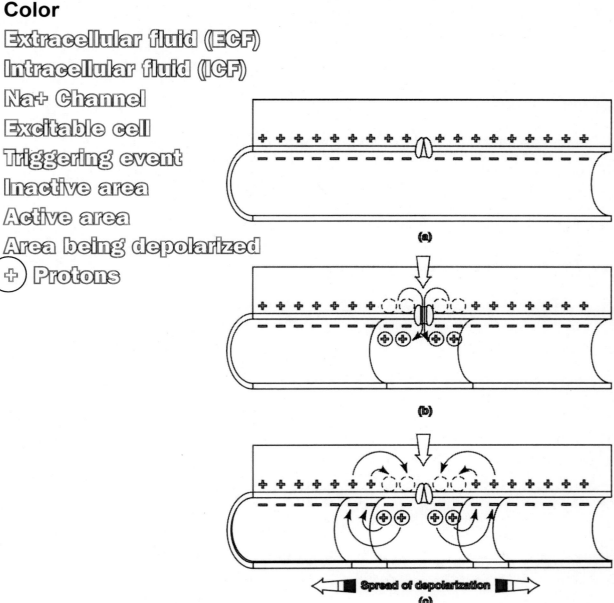

4-5 Conformations of voltage-gated sodium and potassium channels.

Color

Extracellular fluid (ECF)

Intracellular fluid (ICF)

Plasma membrane

Activation gate

Inactivation gate

Potassium

Sodium

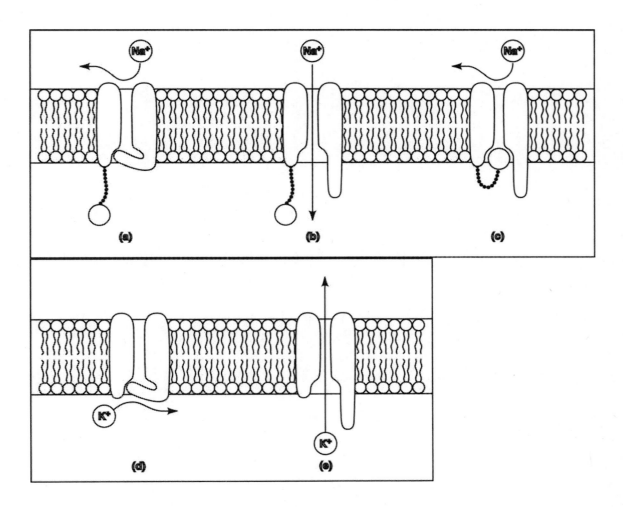

Figure 4-6 Permeability changes and ion fluxes during an action potential.

Voltage-gated Na+ channels
Voltage-gated K+ channels

Take these steps and match them to the numbers in the figure:
___ Further outward movement of K+ through still-open K+ channel briefly hyperpolarizes membrane, which generates after hyperpolarization.
___ At peak of action potential, Na+ inactivation gate closes and P_{Na+} falls, ending net movement of Na+ into cell. At the same time, K+ activation gate opens and P_{K+} rises.
___ At threshold, Na+ activation gate opens and P_{Na+} rises.
___ K+ activation gate closes, and membrane returns to resting potential.
___ Resting potential: all voltage-gated channels closed.
___ K+ leaves cell, causing its repolarization to resting potential, which generates falling phase of action potential.
___ Na+ enters cell, causing explosive depolarization to +30 mV, which generates rising phase of action potential.
___ On return to resting potential, Na+ activation gate closes and inactivation gate opens, resetting channel to respond to another depolarizing triggering event.

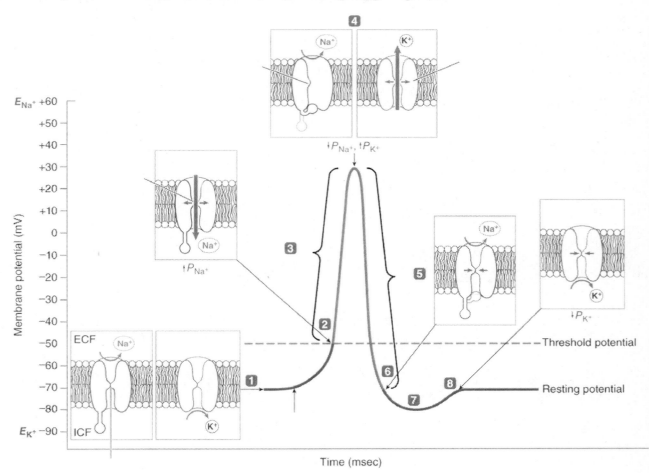

Time (msec)

Figure 4-8 Anatomy of most common type of neuron.

Color	Label
Nucleus	Input zone
Dendrites	Trigger zone
Cell body	Conducting zone
Axon hillock	Output zone
Axon	
Axon terminals	

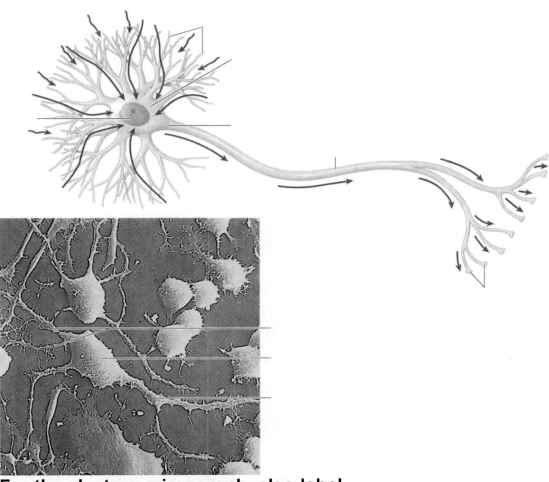

For the electron micrograph, also label

Dendrites

Cell body

Axon

4-9 Contiguous conduction.

Local current flow between the active area at the peak of an action potential and the adjacent inactive area still at resting potential reduces the potential in this contiguous inactive area to threshold, which triggers an action potential in the previously inactive area. The original active area returns to resting potential, and the new active area induces an action potential in the next adjacent inactive area by local current flow as the cycle repeats itself down the length of the axon.

Color and Label

Active at peak of
   action potential

Where depolarization is
   spreading

Graded potential

Sodium

Potassium

Nucleus

Axon terminals

Cell body

Dendrites

Axon at resting
   potential

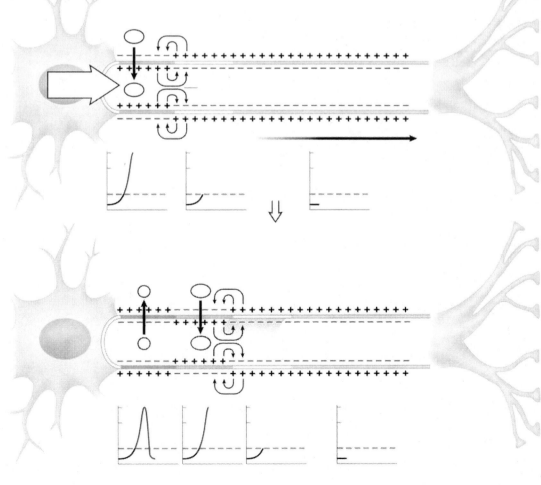

4-12 Saltatory conduction.

The impulse "jumps" from node to node in a myelinated fiber.

Color

Na+

K+

Myelin sheath

Axon

Direction of propagation
   of action potential

Voltage-gated

Na+ and K+ channels

Node of Ranvier

Label what is happening
at every stage

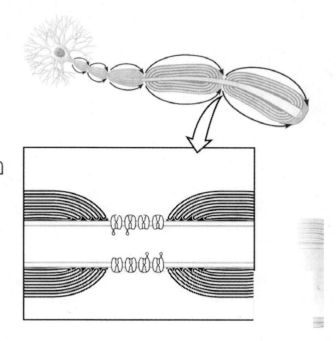

4-15 Synaptic structure and function.

(a) Schematic representation of the structure of a single synapse. The circled numbers designate the sequence of events that take place at a synapse.

Color

Synaptic knob

Axon of presynaptic neuron

Synaptic vesicle

Synaptic cleft

Subsynaptic membrane

Postsynaptic neuron

Voltage-gated Ca2+ channel

Ca2+ Neurotransmitter

Chemically gated ion channel for Na+, K+, or Cl-

Receptor for neurotransmitter

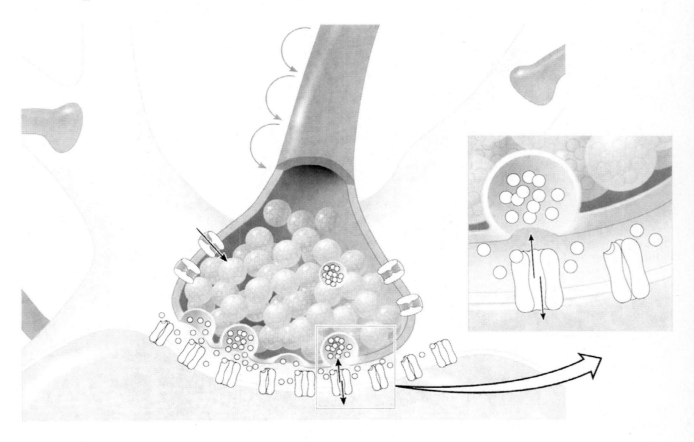

4-16 Determination of the grand postsynaptic potential by the sum of activity in the presynaptic inputs.

Two excitatory (Ex1 and Ex2) and one inhibitory (In1) presynaptic inputs terminate on this hypothetical postsynaptic neuron. The potential of the postsynaptic neuron is being recorded.

Color and label

Excitatory presynaptic input (EX1)

Excitatory presynaptic input (EX2)

Inhibitory (In1)

Postsynaptic cell

Panel A

Panel B

Panel C

Panel D

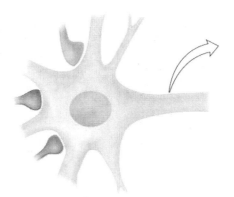
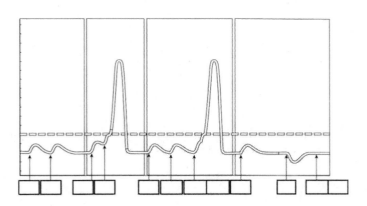

Explain what is happening in each panel of the graph

Panel A _____

Panel B _____

Panel C _____

Panel D _____

4-17 Presynaptic inhibition.

A, an excitatory terminal ending on postsynaptic cell C, is itself innervated by inhibitory terminal B. Stimulation of terminal A alone produces an EPSP in cell C, but simultaneous stimulation of terminal B prevents the release of excitatory neurotransmitter from terminal A. Consequently, no EPSP is produced in cell C despite the fact that terminal A has been stimulated. Such presynaptic inhibition selectively depresses activity from terminal A without suppressing any other excitatory input to cell C. Stimulation of excitatory terminal D produces an EPSP in cell C even though inhibitory terminal B is simultaneously stimulated because terminal B only inhibits terminal A.

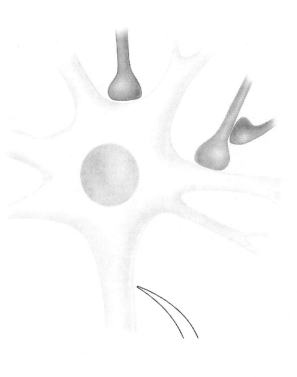

Color

Postsynaptic cell

Excitatory terminal (A)

Inhibitory terminal (B)

Excitatory terminal (D)

Color and Label

Excitatory terminal (A)

Inhibitory terminal (B)

Excitatory terminal (D)

Threshold potential

Resting potential

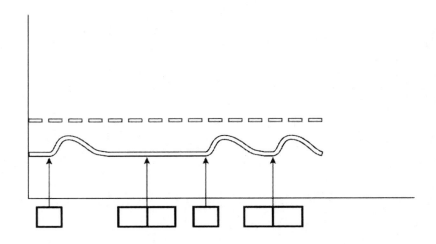

4-19 Types of intercellular communication.

Gap junctions and transient direct linkup of cells by means of complementary surface markers are both means of direct communication between cells. Paracrines, neurotransmitters, hormones, and neurohormones are all extracellular chemical messengers that accomplish indirect communication between cells. These chemical messengers differ in their source and the distance they travel to reach their target cells.

Color

Cell

Nucleus

Neuron

Distant target cell

Secreting cell

Nontarget cell

Blood

Paracrine

Hormone

Neurohormone

Local target cell

Electrical signal

Small molecules and ions

Gap junction

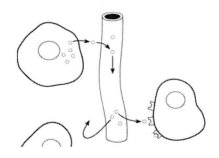

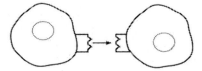

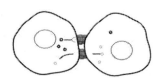

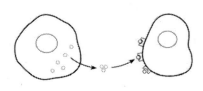

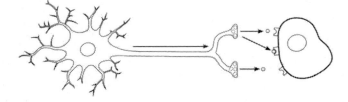

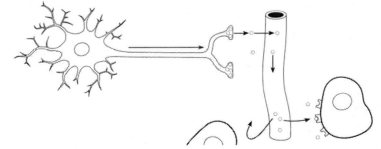

4-25 Mechanism of action of hydrophilic hormones via activation of the cyclic AMP second-messenger pathway.

Color

Extracellular
messenger

ATP

G protein intermediary

G protein receptor

Cyclic AMP

Plasma membrane

Proteins

Extracellular fluid (ECF)

Intracellular fluid (ICF)

Adenylyl cyclase

Label and describe the 5 steps in this process on the figure.

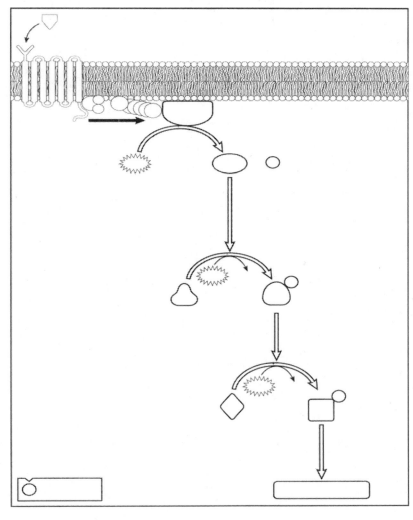

5-3 Glial cells of the central nervous system.

The glial cells include the astrocytes, oligodendrocytes, microglia, and ependymal cells.

Color

Cerebrospinal fluid

Ependymal cell

Neurons

Nucleus

Oligodendrocyte

Astrocyte

Capillary

Microglial cell

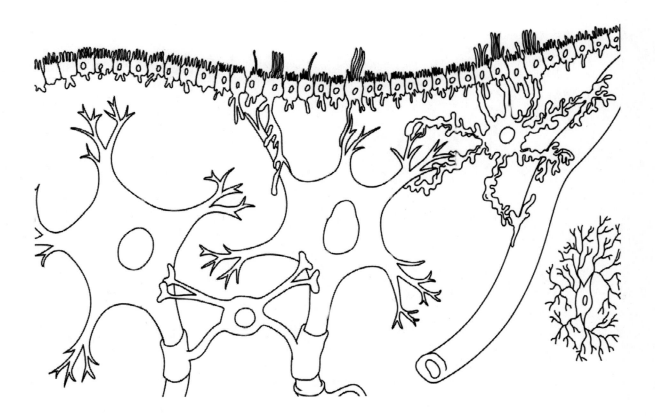

5-5 The ventricles of the brain

Identify

Lateral view of ventricles

- Front of brain
- Back of brain

Anterior view of ventricles

Label

Right lateral ventricle

Left lateral ventricle

Third ventricle

Fourth ventricle

Central canal of spinal cord

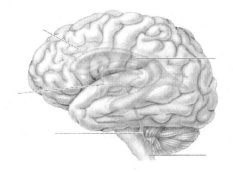

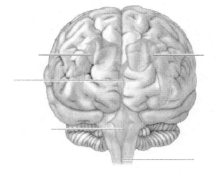

5-6 Relationship of the meninges and cerebrospinal fluid to the brain and spinal cord.

(a) Brain, spinal cord, and meninges in sagittal section. The arrows and numbered steps indicate the direction of flow of cerebrospinal fluid (in yellow). (b) Frontal section in the region between the two cerebral hemispheres of the brain, depicting the meninges in greater detail.

Color

Epithelial tissue
Connective tissue
Muscle tissue
Nervous tissue
Organ
Brain (cerebrum)
Brain stem
Cerebellum
Choroid plexus of fourth
   ventricle
Skull bone
Cerebrospinal fluid
Choroid plexus of third ventricle
Choroid plexus of lateral
   ventricle
Dura mater
Lateral ventricle
Vein
Venous blood
Pia mater
Scalp

Label

Arachnoid mater
Arachnoid villus
Spinal meninges
Cranial meninges
Venous sinus
Dural sinus
Subarachnoid space

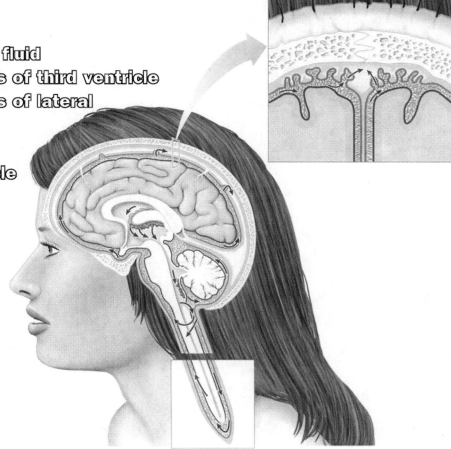

5-2 Overview of the Structures and Functions of the Major Components of the Brain.

Color

Cerebral cortex

Basal nuclei

Thalamus

Hypothalamus

Cerebellum

Mid brain

Pons

Medulla

Spinal Cord

Brain stem (circle)

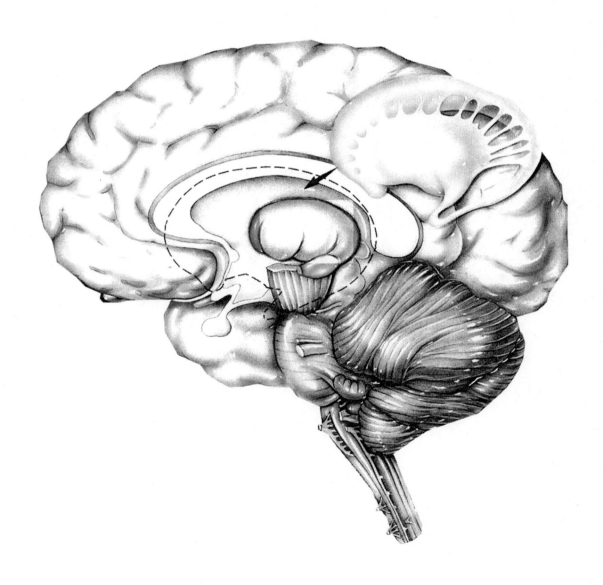

5-12 Functional areas of the cerebral cortex.

(a) Various regions of the cerebral cortex are primarily responsible for various aspects of neural processing, as indicated in this schematic lateral view of the brain.

Color

Primary motor cortex

Somatosensory cortex

Posterior parietal cortex

Wrnicke's area

Primary visual cortex

Premotor

Prefontal association cortex

Broca's area

Brain stem

Spinal cord

Cerebellum

Frontal lobe

Primary auditory cortex

Parietal lobe

Occipital lobe

Temporal lobe

Label

Supplementary motor area

Limbic association cortex

Parietal-temporal-occipital association cortex

Central sulcus

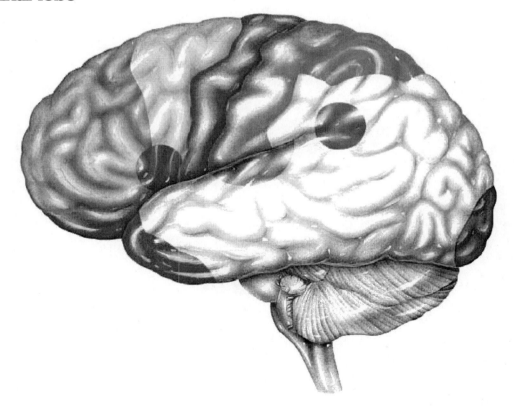

5-17 Possible pathways for long-term potentiation.

Label

Propagation of action potential

EPSP from this source

Other EPSPs

Ca2+ second- messenger
   pathway

Nitric oxide release

Glutamate release

AMPA receptor

NMDA receptor

Color

Presynaptic Neuron

Postsynaptic Neuron

Glutamate

Na+

Ca2+

Mg2+

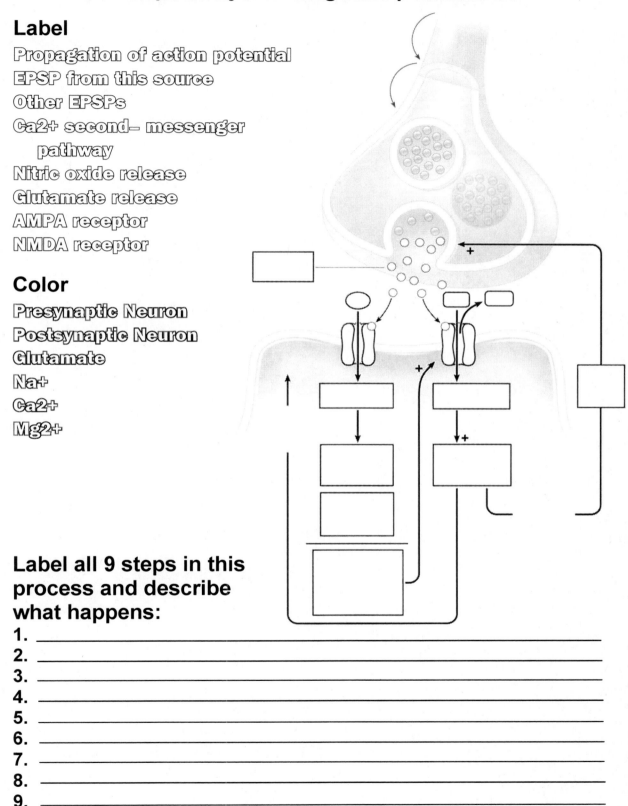

Label all 9 steps in this process and describe what happens:

1. _____

2. _____

3. _____

4. _____

5. _____

6. _____

7. _____

8. _____

9. _____

5-22 Location of the spinal cord relative to the vertebral column.

Color

Spinal cord
Dorsal root ganglion
Meninges
Sympathetic ganglion chain
Vertebra
Spinal nerve
Intervertebral disk

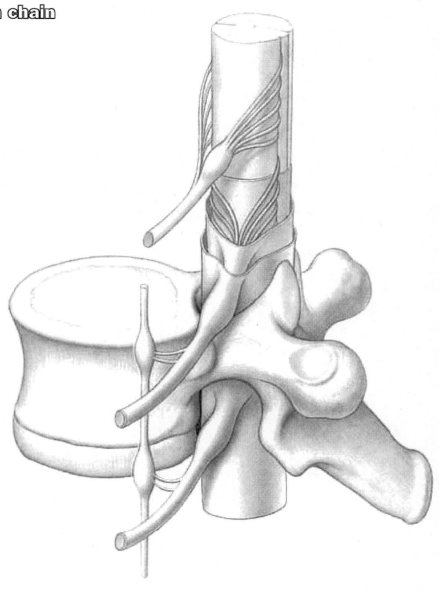

5-24 Spinal cord in cross-section.

Color and Label

White matter
Gray matter
Cell body of efferent neuron
Interneuron
Afferent fiber
Dorsal root
Cell body of afferent neuron

Dorsal root ganglion
Efferent fiber
From receptors
To effectors
Ventral root
Spinal nerve

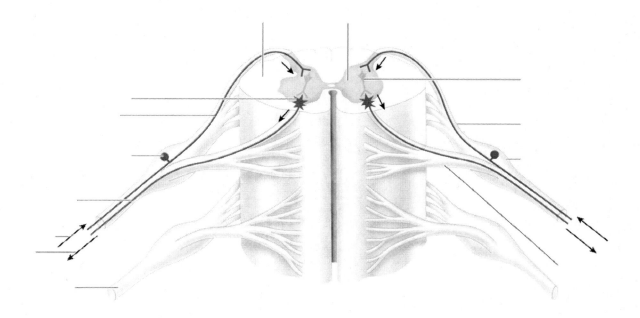

5-28 Structure of a nerve.

Neuronal axons (both afferent and efferent fibers) are bundled together into connective tissue–wrapped fascicles. A nerve consists of a group of fascicles enclosed by a connective tissue covering and following the same pathway. The photograph is a scanning electron micrograph of several nerve fascicles in cross section. (Photos from Dr. R. G. Kessel and Dr. R. H. Kardon/Visuals Unlimited)

Color

Axon

Myelin sheath

Nerve

Connective tissue around the
    nerve

Connective tissue around a
    fascicle

Connective tissue around the
    axon

Blood vessels

Nerve

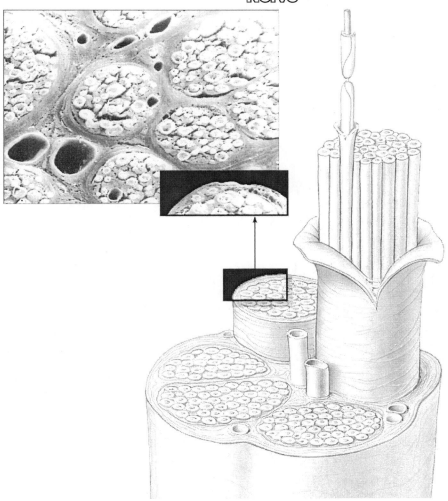

5-29a Distribution of dermatomes

Each of the dermatomes is a skin region with sensory innervation provided by a specific spinal nerve and is designated in the figure by the name of the nerve supplying this area.

Color

Cervical region

Thoracic region

Lumbar region

Sacral region

Label*

C2 through C8

T1 through T12

L1 through L5

S1 through S5

*Color and label both halves of the diagram.

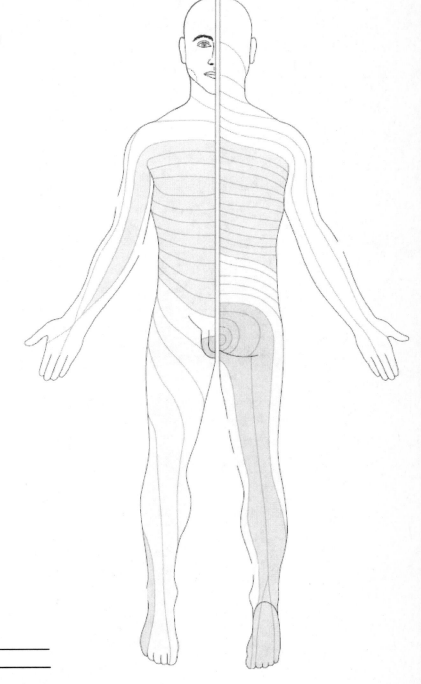

Define "dermatome":

_____

_____

_____

5-29b Shingles rash involving one dermatome

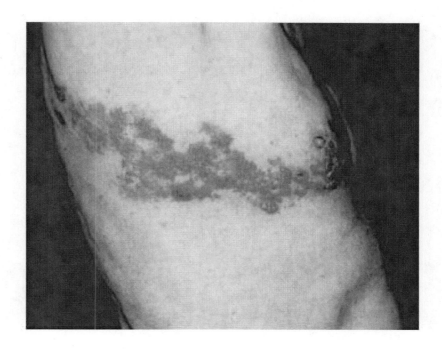

- **Shingles is caused by** _____ **,
  the same virus that causes chicken pox.**

- **Where and how does the rash form?**

  _____
  _____
  _____

- **Shingles is most prevalent in people older than _____
  years old and affects approximately _____ % of
  individuals that have had chicken pox.**

6-1 Conversion of receptor potential into action potential.

(a) Specialized afferent ending as sensory receptor. Local current flow between a depolarized receptor ending undergoing a receptor potential and the adjacent region initiates an action potential in the afferent fiber by opening voltage-gated Na1 channels.
(b) Separate receptor cell as sensory receptor. The depolarized receptor cell undergoing a receptor potential releases a neurotransmitter that binds with chemically gated channels in the afferent fiber ending. This binding leads to a depolarization that opens voltage-gated Na1 channels, initiating an action potential in the afferent fiber.

Color

Stimulus sensitive
    nonspecific cation
    channel

Afferent neuron fiber

Receptor

Voltage-gated Ca²⁺
    channel

Neurotransmitter

Voltage-gated Na⁺
    channel

Chemically gated
    receptor-channel

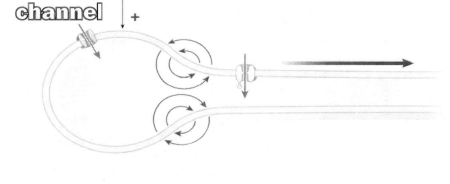

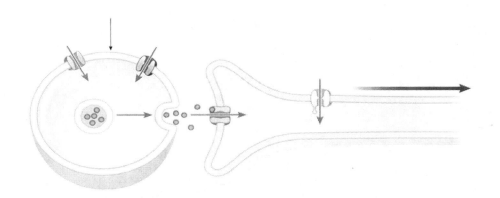

6-1 Conversion of receptor potential into action potential (cont).

Label

Stimulus

Action potential

Chemically generated
   channels

Label all 3 steps in the process depicted in figure (a) and describe what happens:

1. _____

_____

2. _____

_____

3. _____

_____

Label all 6 steps in the process depicted in figure (b) and describe what happens:

1. _____

_____

2. _____

3. _____

_____

4. _____

_____

5. _____

_____

6. _____

_____

6-5 Tactile receptors in the skin.

Color

Subcutaneous tissue

Pacinian corpuscle

Ruffini endings

Meissner's corpuscle

Shaft of hair inside follicle

Skin surface

Myelinated neuron

Epidermis

Dermis

Merkel's disc

Hair receptor

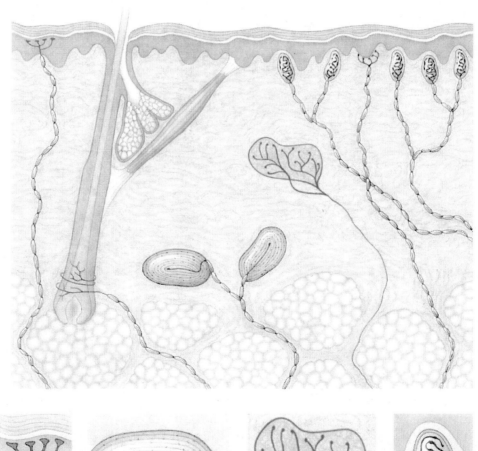

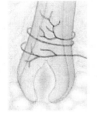

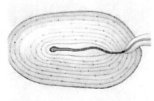

6-6 Comparison of discriminative ability of regions with small verses large receptive fields.

The relative tactile acuity of a given region can be determined by the *two-point threshold-of-discrimination test*. If the two points of a pair of calipers applied to the surface of the skin stimulate two different receptive fields, two separate points are felt. If the two points touch the same receptive field, they are perceived as only one point. By adjusting the distance between the caliper points, one can determine the minimal distance at which the two points can be recognized as two rather than one, which reflects the size of the receptive fields in the region. With this technique, it is possible to plot the discriminative ability of the body surface. The two point threshold ranges from 2mm in the fingertip (enabling a person to read Braille, where the raised dots are spaced 2.5mm apart) to 48mm in the poorly discriminative skin of the calf of the leg.

Color

Calipers

Receptive fields on skins surface

Receptor endings

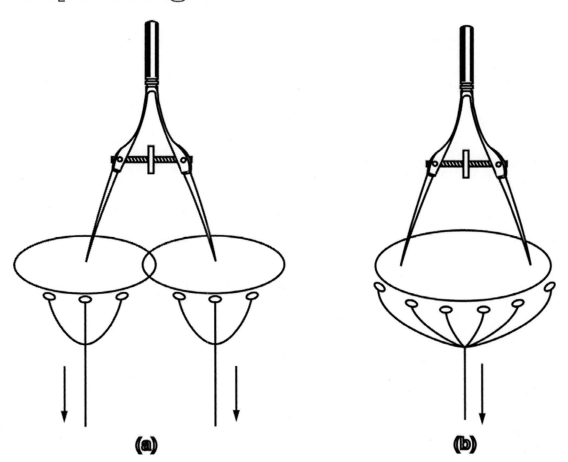

(a)　　　　　(b)

6-7 Lateral inhibition.

(a) The receptor at the site of most intense stimulation is activated to the greatest extent. Surrounding receptors are also stimulated but to a lesser degree. (b) The most intensely activated receptor pathway halts transmission of impulses in the less intensely stimulated pathways through lateral inhibition. This process facilitates localization of the site of stimulation.

Color

Skin surface

Receptor pathways

Area stimulated the most

Area stimulated the least

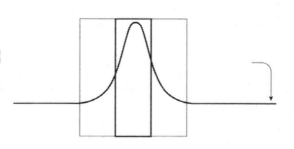

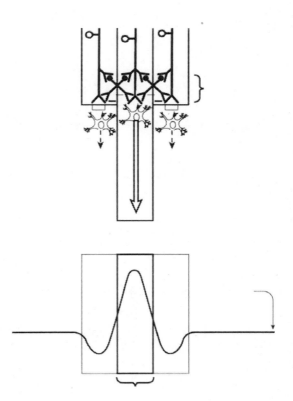

6-9 Substance P pain pathway and analgesic pathway.

(a) Substance P pain pathway. When activated by a noxious stimulus, some afferent pain pathways release substance P, which activates ascending pain pathways that provide various brain regions with input for processing different aspects of the painful experience. (b) Analgesic pathway. Endogenous opiates released from descending analgesic (pain-relieving) pathways bind with opiate receptors at the synaptic knob of the afferent pain fiber. This binding inhibits the release of substance P, thereby blocking transmission of pain impulses along the ascending pain pathways.

Color

Noxious stimulant
Substance P
Endogenous opiate
Neuron
Opiate receptors

Label

Higher brain
Brain stem
Spinal cord
Reticular Formation
Afferent Pain Fiber
Thalamus
Somatosensory cortex
Nociceptor
Hypothalamus
Periaqueductal gray matter

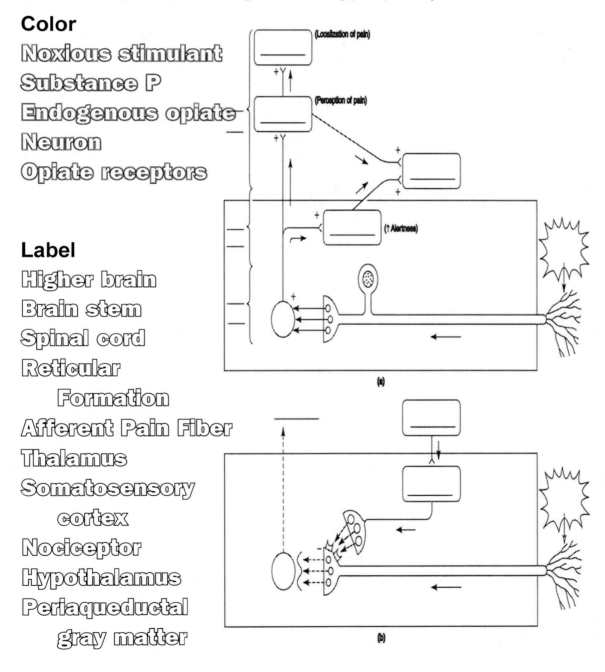

6-10 Structure of the eye.

(a) External front view (b) Internal sagittal view

Color

Lens
Cornea
Ciliary body
Conjunctiva
Iris
Pupil
Retina
Sclera
Optic nerve
Aqueous humor
Vitreous humor
Canal for tear drainage
Sclera

Label

Blood vessels in
    retina
Suspensory
    ligament
Extrinsic eye
    muscle
Choroid
Fovea
Optic disc

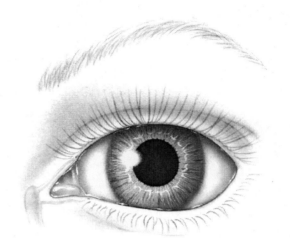

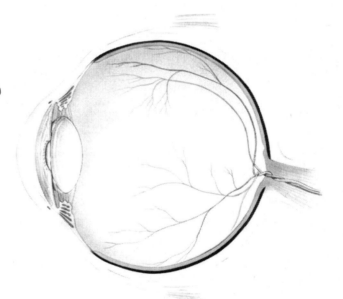

127

6-12 Properties of an electromagnetic wave

A wavelength is the distance between two wave peaks. The intensity is the amplitude of the wave.

Label

 One

wavelength

Distance

Intensity

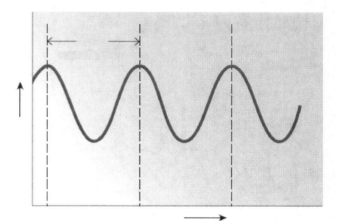

In reference to color sensations, the shorter visible wavelengths are sensed as _____and _____; the longer wavelengths are interpreted as _____ and _____.

T/F   Visible light accounts for all different wavelengths.

T/F The act of dimming a bright blue light changes its color.

T/F Light waves radiate outward in all directions from every point of a light source.

6-14 Focusing of diverging light rays.

Diverging light rays must be bent inward to be focused.

Color

Point source of light

Light rays

Light rays focused on the retina

Retina

Cornea

Lens

Aqueous humor

Vitreous humor

Ciliary body

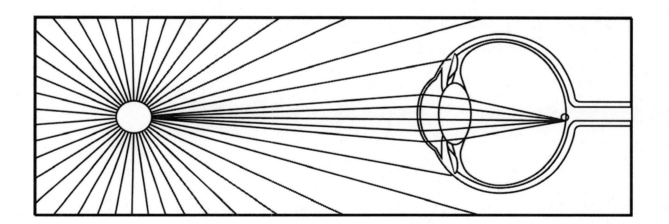

6-20 Emmetropia, myopia, and hyperopia.

This figure compares far vision and near vision (a) in the normal eye with (b) nearsightedness and (c) farsightedness in both their (1) uncorrected and (2) corrected states. The vertical dashed line represents the normal distance of the retina from the cornea; that is, the site at which an image is brought into focus by the refractive structures in a normal eye.

Color

Lens

Corrective lens

Eye

List the ability to see for near and far source sight and why these problems may happen:

1. Normal eye

_____

_____

_____

_____

_____

2. Nearsightedness

_____

_____

_____

_____

_____

_____

3. Farsightedness

_____

_____

_____

_____

_____

_____

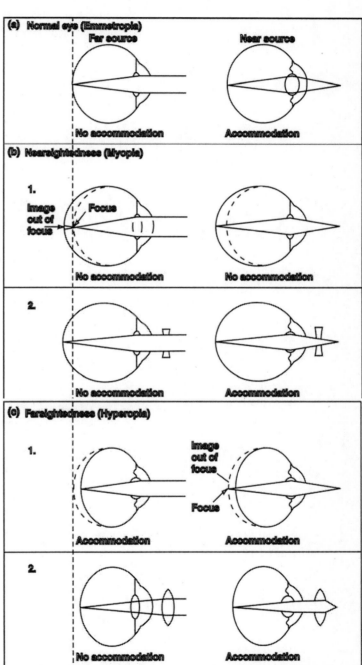

6-22 Retinal layers.

The retinal visual pathway extends from the photoreceptor cells (rods and cones, whose light-sensitive ends face the choroid away from the incoming light) to the bipolar cells to the ganglion cells. The horizontal and amacrine cells act locally for retinal processing of visual input.

Color

Fibers of the optic nerve

Ganglion cell

Amacrine cell

Bipolar cell

Horizontal cell

Cone

Photoreceptor cells

Rod

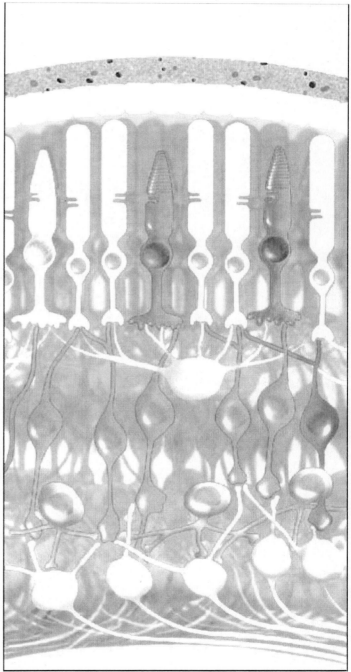

Figure 6-30 The visual pathway and visual deficits associate with lesions in the pathway.

Color and Label
Optic nerve
Optic chiasm
Optic tract

Thalamus
Optic radiation
Visual cortex

Location of overlying frontal lobes

Right eye

Left eye

1

2

3

Viewing brain from above with overlying structures removed

(a) Visual pathway

Site of lesion	Visual deficit	
	Left eye	Right eye
1 Left optic nerve		
2 Optic chiasm		
3 Left optic tract (or radiation)		

(b) Visual deficits with specific lesions in visual pathway

For (b)—cross off with an X the sites of visual deficits associated with each lesion.

6-35 Transmission of sound waves.

(a) Fluid movement within the cochlea set up by vibration of the oval window follows two pathways, one dissipating sound energy and the other initiating the receptor potential.

Color

Outer ear
Tympanic membrane
Basilar membrane
Tectorial membrane
Hairs
Organ of corti
Malleus
Stapes
Incus
Cochlea
Scala tympani
Scala vestibule

Label

Helicotrema
Perilymph
Round window
Oval window

Describe what happened in both pathways what happens:

1. _____

_____

_____

2. _____

_____

_____

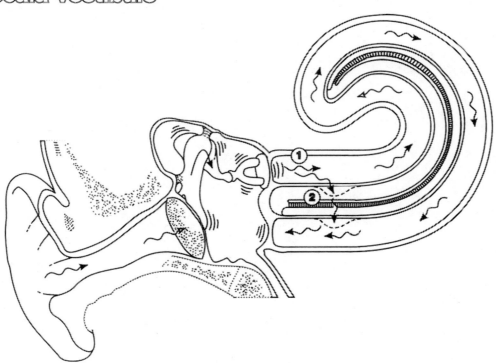

139

6-36 Bending of hairs on deflection of the basilar membrane.

The stereocilia (hairs) from the hair cells of the basilar membrane contact the overlying tectorial membrane. These hairs are bent when the basilar membrane is deflected in relation to the stationary tectorial membrane. This bending of the inner hair cells' hairs opens mechanically gated channels, leading to ion movements that result in a receptor potential.

Color

Hair cells

Tectorial membrane

Basilar membrane

Area of fluid movement

Area of ion movement resulting in receptor
    potential

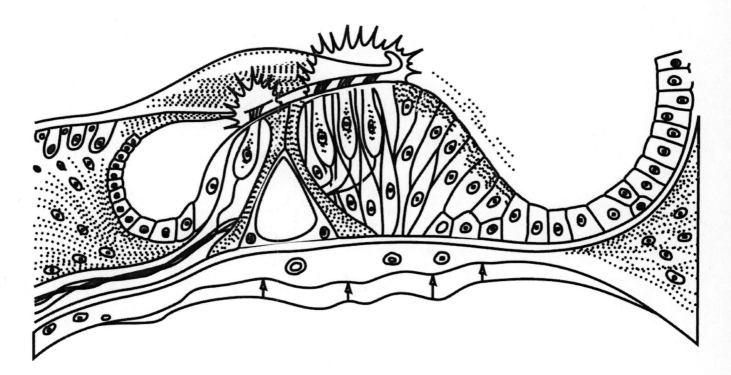

6-44 Location and structure of the olfactory receptor cells.

Color
Olfactory bulb
Afferent nerve fiber A
Afferent nerve fiber B
Afferent nerve fiber C
Basal cell
Mucus layer
Olfactory receptor cell A
Olfactory receptor cell B
Olfactory receptor cell C
Olfactory mucosa
Supporting cell
Cilia
Bone

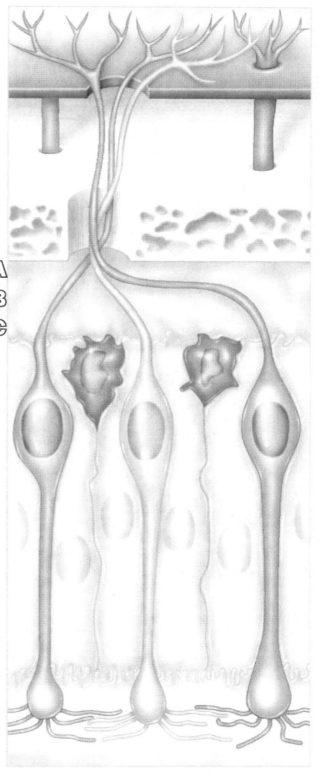

143

7-1 Autonomic nerve pathway.

Color

Central nervous system
Effector organ
Varicosity
Preganglionic neurotransmitter
Preganglionic fiber
Autonomic ganglion
Postganglionic neurotransmitter
Nucleus

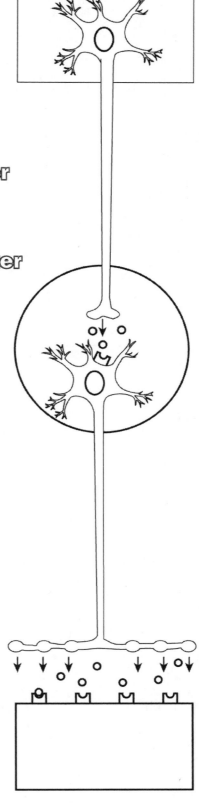

7-2 Autonomic nervous system.

The sympathetic nervous system, which originates in the thoracolumbar regions of the spinal cord, has short cholinergic (acetylcholine-releasing) preganglionic fibers and long adrenergic (norepinephrine-releasing) postganglionic fibers. The parasympathetic nervous system, which originates in the brain and sacral region of the spinal cord, has long cholinergic preganglionic fibers and short cholinergic postganglionic fibers. In most instances, sympathetic and parasympathetic postganglionic fibers both innervate the same effector organs. The adrenal medulla is a modified sympathetic ganglion, which releases epinephrine and norepinephrine into the blood. Nicotinic cholinergic receptors are located in the autonomic ganglia and adrenal medulla and respond to ACh released by all autonomic preganglionic fibers. Muscarinic cholinergic receptors are located at the autonomic effectors and respond to ACh released by parasympathetic postganglionic fibers. a1, a2, b1, b2 adrenergic receptors are variably located at the autonomic effectors and differentially respond to norepinephrine released by sympathetic postganglionic fibers and to epinephrine released by the adrenal medulla.

Color

Brain

Spinal cord

Adrenal medulla

Blood vessel

Blood

Effector organs

Sympathetic ganglion chain

Collateral ganglion

Terminal ganglion

Acetylcholine

Norepinephrine

Epinephrine

β_1 receptor

β_2 receptor

α receptor

Nicotinic receptor

Muscarinic receptor

Label

Preganglionic fiber

Postganglionic fiber

Sympathetic preganglionic fiber

Sympathetic postganglionic fiber

Cardiac Muscle

Smooth Muscle

Endocrine glands

Adipose tissue

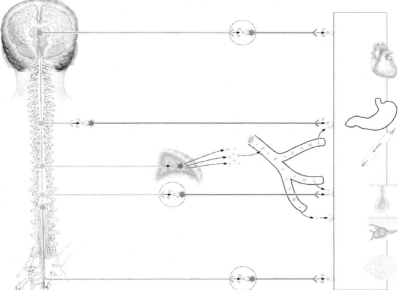

7-5 Events at a neuromuscular junction.

Color

Axon terminal of motor
    neuron
Myelin sheath
Acetylcholinesterase
Acetylcholine-gated receptor-
    channel
Voltage-gated calcium
    channel
Contractile elements within
    muscle fiber
Terminal button
Vesicle of acetylcholine

Voltage-gated sodium channel
Plasma membrane of muscle
    fiber
Sodium
Potassium

Label

Motor end plate
Action potential propagation in
    motor neuron
Action potential
    propagation in muscle
    fiber

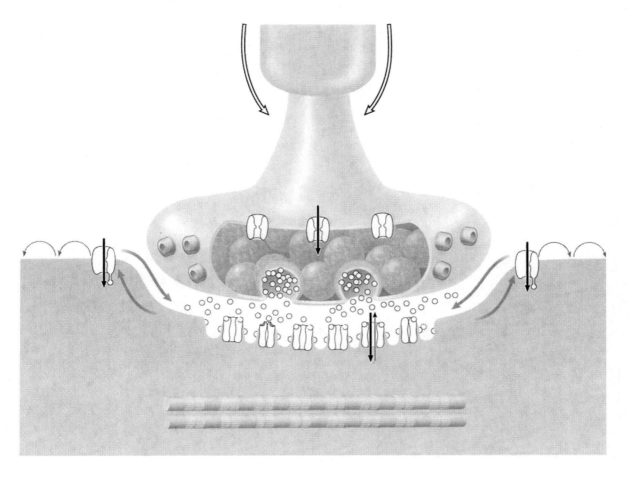

8-2 Levels of organization in a skeletal muscle.

Note in the cross section of a myofibril in part (c) that each thick filament is surrounded by six thin filaments, and each thin filament is surrounded by three thick filaments.

Color

Z line
Thin filaments
Thick filaments
Cross bridge
I band
A band
Muscle
Tendon
Muscle fiber
Myofibril
Actin
Troponin
Tropomyosin
Myosin head
Myosin tail

Label

H Zone
M line
I band
Sarcomere
A band

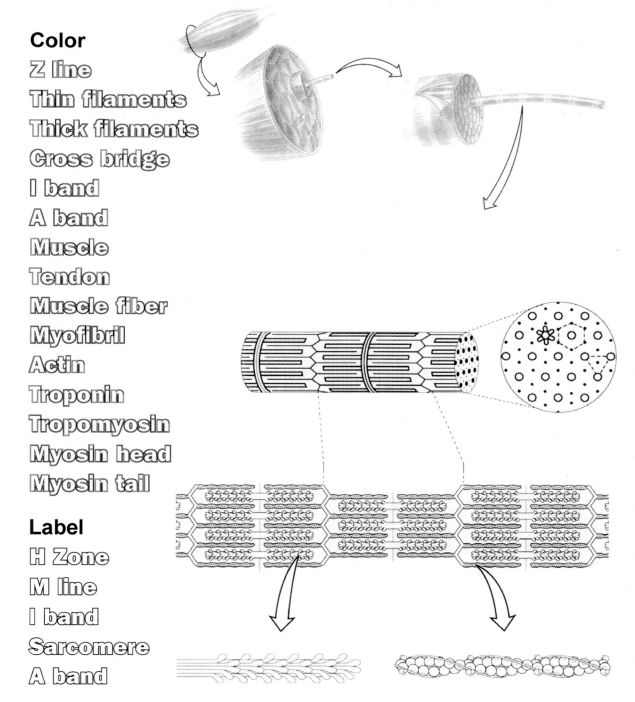

8-4 Structure of myosin molecules and their organization within a thick filament.

(a) Each myosin molecule consists of two identical, golf-club-shaped subunits with their tails intertwined and their globular heads, each of which contains an actin binding site and a myosin ATPase site, projecting out at one end. (b) A thick filament is made up of myosin molecules lying lengthwise parallel to one another. Half are oriented in one direction and half in the opposite direction. The globular heads, which protrude at regular intervals along the thick filament, form the cross bridges.

Color

Head A

Head B

Tail A

Tail B

Actin-binding site

Myosin ATPase site

Cross bridges

Myosin molecules

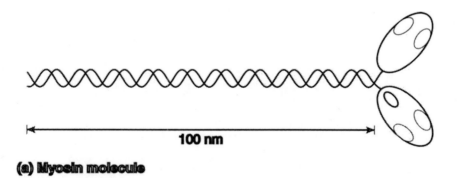

100 nm

(a) Myosin molecule

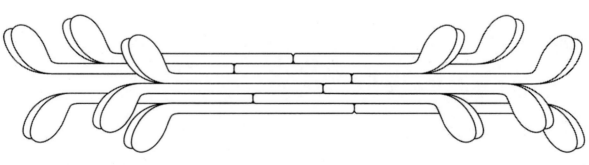

(b) Thick filament

8-5 Composition of a thin filament.

The main structural component of a thin filament is two chains of spherical actin molecules that are twisted together. Troponin molecules (which consist of three small, spherical subunits) and threadlike tropomyosin molecules are arranged to form a ribbon that lies alongside the groove of the actin helix and physically covers the binding sites on actin molecules for attachment with myosin cross bridges. (The thin filaments shown here are not drawn in proportion to the thick filaments in Figure 8-4. Thick filaments are two to three times larger in diameter than thin filaments.)

Color

Binding site

Actin Molecules

Tropomysin

Troponin

Label

Actin Helix

Thin Filament

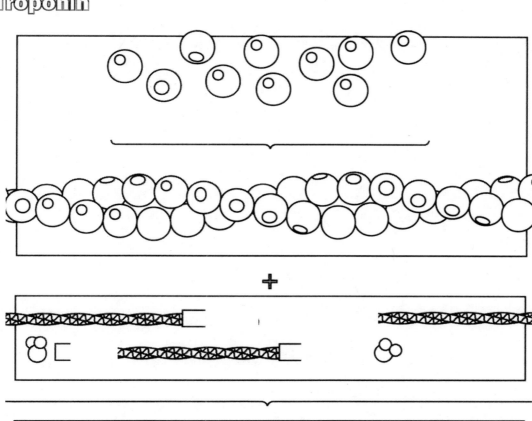

8-6 Role of calcium in turning on cross bridges.

Color

Actin binding site

Myosin binding site

Myosin

Actin molecules

Tropomyosin

Troponin

Describe the 3 stages of a relaxed cross bridge, and the 4 stages of an excited cross bridge.

1._____

2._____

3._____

1._____

2._____

3._____

4._____

8-7 Changes in banding pattern during shortening.

During muscle contraction, each sarcomere shortens as the thin filaments slide closer together between the thick filaments so that the Z lines are pulled closer together. The width of the A bands does not change as a muscle fiber shortens, but the I bands and H zones become shorter.

Color

Thick filaments

Thin filaments

Sarcomere

H zone

I band

A band

Z line

Troponin

Label

Sarcomere

H zone

I band

A band

Z line

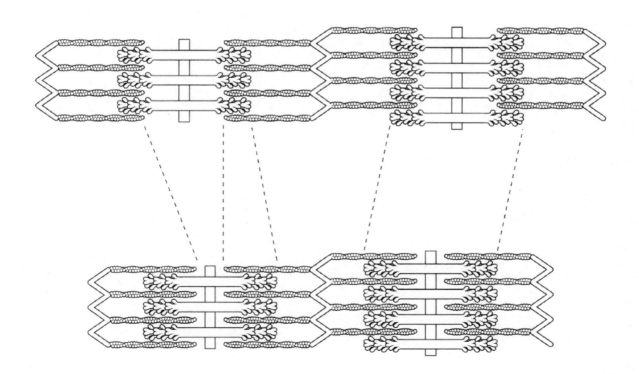

8-8 Cross-bridge activity.

(a) During each cross-bridge cycle, the cross bridge binds with an actin molecule, bends to pull the thin filament inward during the power stroke, then detaches and returns to its resting conformation, ready to repeat the cycle. (b) The power strokes of all cross bridges extending from a thick filament are directed toward the center of the thick filament. (c) All six thick filament is surrounded on each end by six thin filaments, all of which are pulled inward simultaneously through cross-bridge cycling during muscle contraction.

Color

Z line

Actin molecules

Myosin Cross bridge

Thin myofilament

Thick myofilament

Label all 4 steps in the cross bridge cycle and describe what happens:

1. _____

_____

_____

2. _____

_____

_____

3. _____

_____

_____

4. _____

_____

_____

8-11 Excitation–contraction coupling and muscle relaxation.

Steps **1** through **5** show the events that couple neurotransmitter release and subsequent electrical excitation of the muscle cell with muscle contraction. Steps **6** and **7** show events associated with muscle relaxation.

Color

Terminal button

Acetylcholine

Tropomyosin

T tubule

Acetylecholine-gates
   cation channel

Surface membrane of
   muscle cell

Lateral sacs of
   sarcoplasmic reticulum

Troponin

Actin

Myosin

Myosin cross bridge

Cross bridge binding site

Calcium

Label steps 1-7

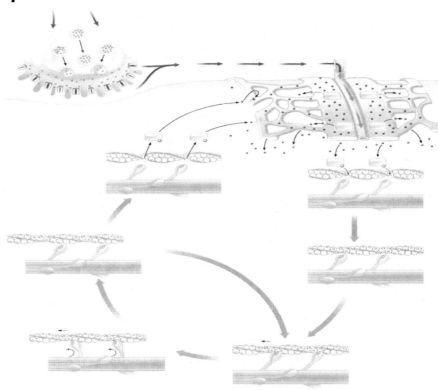

8-12 Cross-bridge cycle.

Color

Actin

Myosin

Label all 4 steps of the cross-bridge cycle and describe each step

1. _____

2a. _____

2b. _____

3. _____

4a. _____

4b. _____

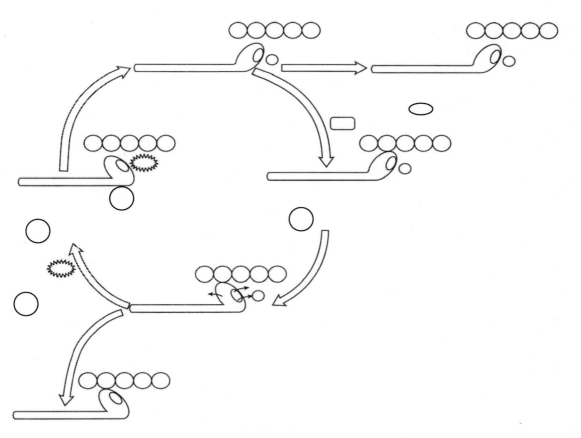

8-17 Lever systems of muscles, bones, and joints.

Note that the lever ratio (length of the power arm to length of the load arm) is 1:7 (5 cm:35 cm), which amplifies the distance and velocity of movement seven times (distance moved by the muscle [extent of shortening] 5 1 cm, distance moved by the hand 5 7 cm, velocity of muscle shortening 5 1 cm/unit of time, hand velocity 5 7 cm/unit of time), but at the expense of the muscle having to exert seven times the force of the load (muscle force 5 35 kg, load 5 5 kg).

Color

Load

Fulcrum

Force

Hand/Arm

Radius

Ulna

Scapula

Humerus

Bicep

Label

Upward force

Downward force

Power arm of lever

Load arm of lever

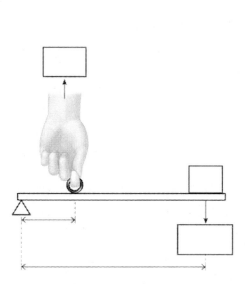

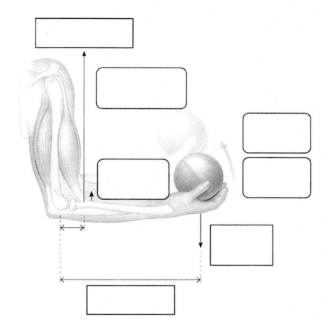

8-23 Muscle receptors.

(a) A muscle spindle consists of a collection of specialized intrafusal fibers that lie within a connective tissue capsule parallel to the ordinary extrafusal skeletal muscle fibers. The muscle spindle is innervated by its own gamma motor neuron and is supplied by two types of afferent sensory terminals, the primary (annulospiral) endings and the secondary (flower-spray) endings, both of which are activated by stretch. (b) The Golgi tendon organ is entwined with the collagen fibers in a tendon and monitors changes in muscle tension transmitted to the tendon.

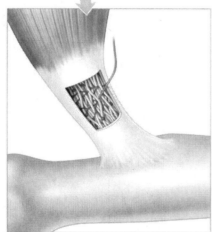

Color

Alpha motor neuron axon

Gamma motor neuron axon

Afferent neuron axons

Secondary endings of
    afferent fibers

Extrafusal muscle fibers

Capsule

Intrafusal muscle fibers

Bone

Skeletal muscle

Tendon

Golgi tendon organ

Collagen

Afferent fiber

Primary endings of afferent
    fibers

Contractile end portions of
    intrafusal fiber

Noncontractile central
    portion of intrafusal fiber

8-25 Patellar tendon reflex (a stretch reflex).

Tapping the patellar tendon with a rubber mallet stretches the muscle spindles in the quadriceps femoris muscle. The resultant monosynaptic stretch reflex results in contraction of this extensor muscle, causing the characteristic knee-jerk response.

Color

Patellar tendon

Extensor muscle of
   knee

Femur

Tibia

Fibula

Muscle spindle

Alpha motor neuron

Hammer

Integrating center

Motion arrows

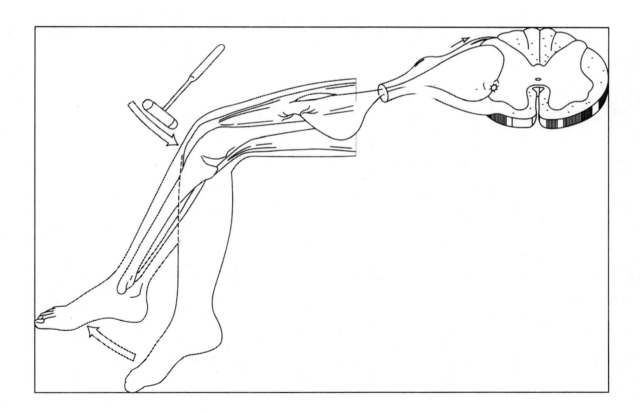

Figure 8-29 Arrangement of thick and thin filaments in a smooth muscle cell in relaxed and contracted states.

Color and Label

Dense body

Bundle of thick and thin filaments

One relaxed contractile unit extending from side to
    side

One contracted contractile unit

Plasma membrane

Thin filament

Thick filament

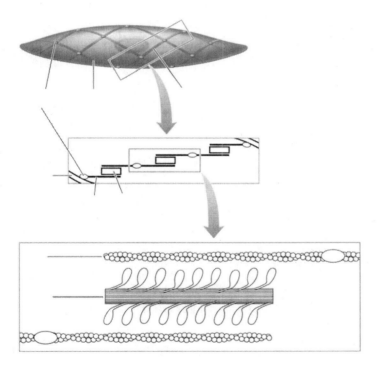

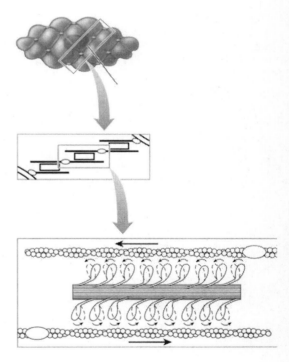

9-1 Pulmonary and systemic circulation in relation to the heart.

The circulatory system consists of two separate vascular loops: the pulmonary circulation, which carries blood between the heart and lungs; and the systemic circulation, which carries blood between the heart and organ systems.

Color

Oxygen rich blood

Oxygen poor blood

Lungs

Heart

Label

Aorta

Capillary networks
   of left lung

Capillary
      networks of
      lower body

Capillary
      networks of
      right lung

Capillary networks
   of upper body

Pulmonary
      arteries

Pulmonary
      veins

Systemic
      circulation

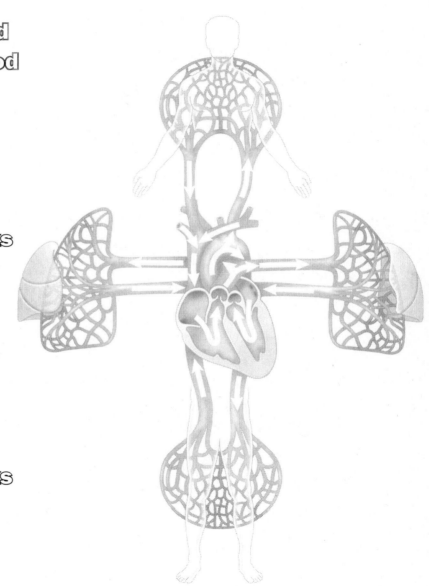

9-2a Blood flow through and pump action of the heart.

(a) Blood flow through the heart. The arrows indicate the direction of blood flow. To illustrate the direction of blood flow through the heart, all of the heart valves are shown open, which is never the case. The right side of the heart receives O2-poor blood from the systemic circulation and pumps it into the pulmonary circulation. The left side of the heart receives O2-rich blood from the pulmonary circulation and pumps it into the systemic circulation.

Color

Arrows indicating O$_2$ rich
   blood

Arrows indicating O$_2$ poor
blood

Right atrium

Right ventricle

Left atrium

Left ventricle

Aorta

Right pulmonary vein

Left pulmonary vein

Superior vena cava

Right pulmonary artery

Left pulmonary artery

Aortic semilunar valve

Interventricular septum

Right atrioventricular (AV)
valve

Left atrioventricular (AV)
valve

Pulmonary semilunar valve

Inferior vena cava

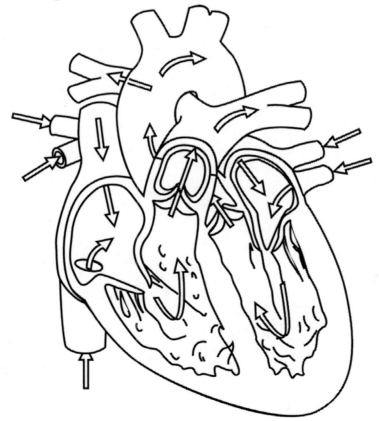

9-2b Blood flow through and pump action of the heart.

Dual pump action of the heart. The right side of the heart receives O2-poor blood from the systemic circulation and pumps it into the pulmonary circulation. The left side of the heart receives O2-rich blood from the pulmonary circulation and pumps it into the systemic circulation. Note the parallel pathways of blood flow through the systemic organs. (The relative volume of blood flowing through each organ is not drawn to scale.)

Color

Other systemic organs

Brain

Digestive tract

Kidneys

Muscles

Lungs

O_2 rich blood

O_2 poor blood

Right atrium and ventricle

Left atrium and ventricle

Label

Venae cavae

Aorta

Pulmonary artery

Pulmonary veins

Systemic circulation

Pulmonary circulation

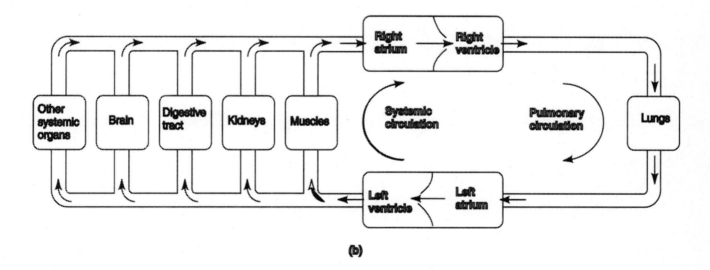

(b)

9-3 Mechanism of valve action.

When pressure is greater behind the valve, it opens. When pressure is greater in front of the valve, it closes. Note that when pressure is greater in front of the valve, it does not open in the opposite direction: that is, it is a one-way valve.

Color

Valve

Arrows from pressure

Vessel wall

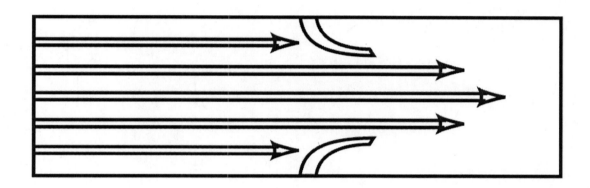

Valve opened

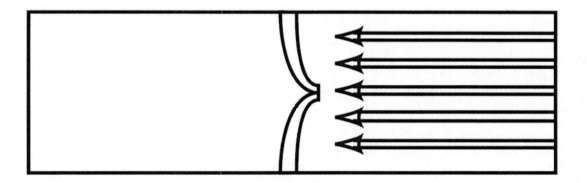

Valve closed; does not open in opposite direction

9-8 Specialized conduction system of the heart.

Color

Right atrium

Left atrium

Right ventricle

Left ventricle

Purkinje fibers

Atrioventricular (AV) node

Interatrial pathway

Left branch of bundle of His

Right branch of bundle of His

Electrically nonconductive fibrous tissue

Sinoatrial (SA) node

Internodal pathway

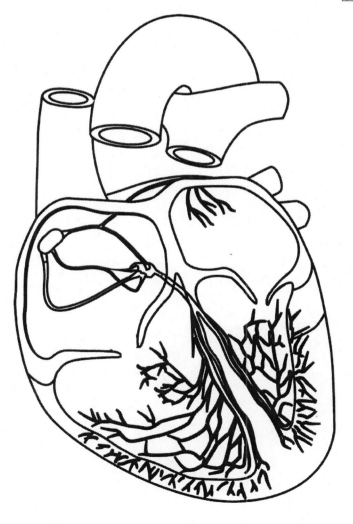

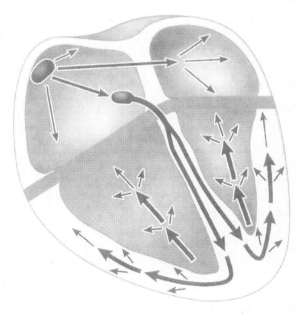

9-16 Cardiac cycle.

The sketches of the heart illustrate the flow of O2-poor (dark blue) and O2-rich (bright red) blood in and out of the ventricles during the cardiac cycle.

Color

Area of O$_2$ right blood

Area of O$_2$ poor blood

Right atrium

Left atrium

Left ventricle

Right ventricle

Passive filling during ventricular and atrial diastole

 A

Atrial contraction

B

Ventricular filling

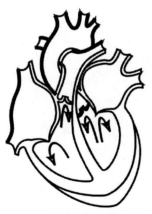

Isovolumetric ventricular contraction

 C

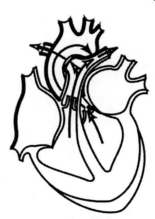

Ventricular ejection

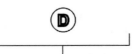 D

Ventricular emptying

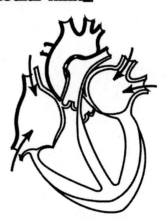

Isovolumetric ventricular relaxation

 E

9-27 Atherosclerotic plaque in a coronary vessel.

Color

Plaque

Endothelium

Lipid-rich core of plaque

Normal blood vessel wall

Collagen-rich smooth muscle cap of plaque

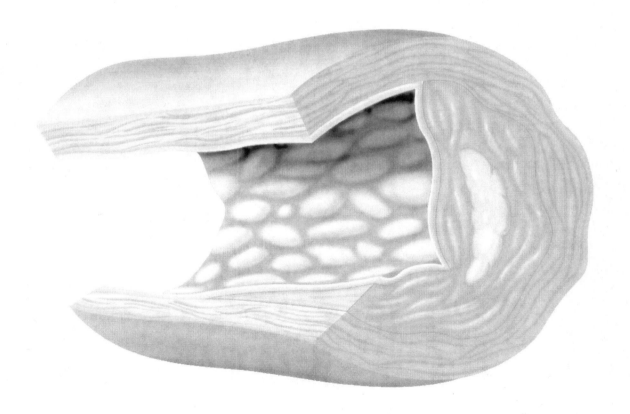

9-28 Consequences of thromboembolism.

(a) A thrombus may enlarge gradually until it completely occludes the vessel at that site.
(b) A thrombus may break loose from its attachment, forming an embolus that may completely occlude a smaller vessel downstream.

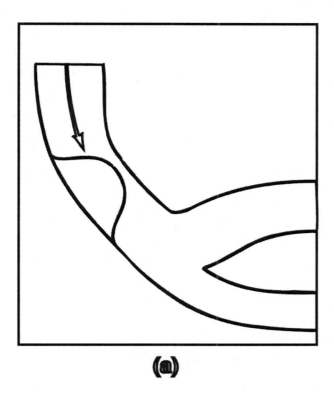

(a)

Color
Bloodflow
Thrombus
Embolus

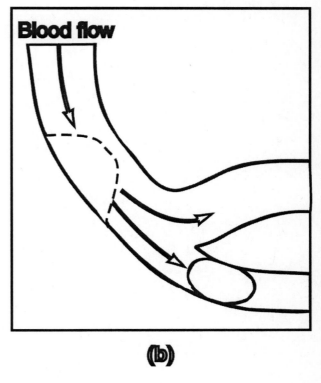

Blood flow

(b)

Table 10-1 Features of Blood Vessels.

Color

Endothelium
Venous valve
Basement membrane

Elastic fibers
Smooth muscle
Connective tissue coat

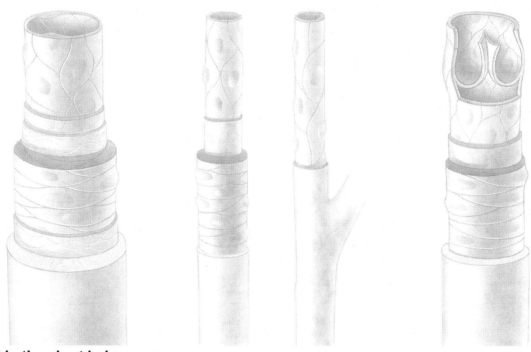

Fill in the chart below

Feature	Arteries	Arterioles	Vessel Type Capillaries	Veins
Number				
Special Features				
Functions				

10-4 Basic organization of the cardiovascular system

Arteries progressively branch as they carry blood from the heart to the organs. A separate small arterial branch delivers blood to each of the various organs. As a small artery enters the organ it is supplying, it branches into arterioles, which further branch into an extensive network of capillaries. The capillaries rejoin to form venules, which further unite to form small veins that leave the organ.

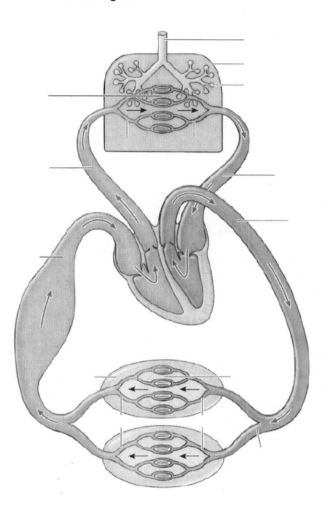

For simplicity, only two capillary beds within two organs are illustrated.

Identify
Pulmonary circulation
Systemic circulation

Color
Circulation to the heart
Circulation away from the heart

Label
Air sac
Airway
Aorta
Arterioles
Lungs
Pulmonary artery
Pulmonary veins
Smaller arteries branching off to supply various tissues
Systemic capillaries

The vascular tree consists of _____, _____, _____, _____, and _____.

10-6 Arteries as a pressure reservoir.

Because of their elasticity, arteries act as a pressure reservoir. (a) The elastic arteries distend during cardiac systole as more blood is ejected into them than drains off into the narrow, high-resistance arterioles downstream. (b) The elastic recoil of arteries during cardiac diastole continues driving the blood forward when the heart is not pumping.

Color

From veins

Heart

Arteries

Expansion area

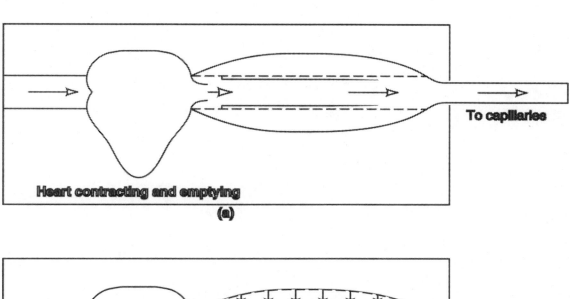

Heart contracting and emptying

To capillaries

(a)

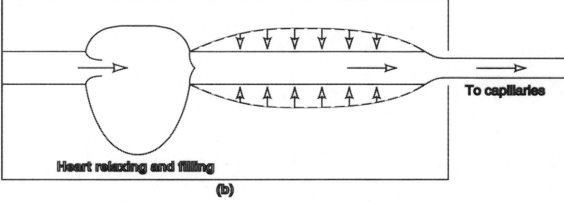

Heart relaxing and filling

To capillaries

(b)

10-9 Arteriolar vasoconstriction and vasodilation.

(b) Schematic representation of an arteriole in cross section showing normal arteriolar tone. (c) Outcome of and factors causing arteriolar vasoconstriction. (d) Outcome of and factors causing arteriolar vasodilation.

Color

Blood

Smooth muscle

Exterior of vessel

Normal arteriolar tone

Cross section of arteriole

(b)

Causes of Vasoconstriction:

_____
_____
_____
_____
_____
_____ **Vasoconstriction**
_____
_____
_____
_____
_____

(c)

Causes of vasodilatation:

_____
_____ **Vasodilation**
_____
_____
_____
_____
_____
_____
_____

(d)

10-17 Exchanges across the capillary wall, the most common type of capillary.

(b) As depicted in this schematic representation of a cross section of a capillary wall, small water-soluble substances are exchanged between the plasma and the interstitial fluid by passing through the water-filled pores, whereas lipid-soluble substances are exchanged across the capillary wall by passing through the endothelial cells. Proteins to be moved across are exchanged by vesicular transport. Plasma proteins generally cannot escape from the plasma across the capillary wall.

Color

Plasma Proteins

Exchange Proteins

Water-filled pore

Endothelial cell – center

Endothelial cell – membrane

Interstitial fluid

Plasma Membrane

Cytoplasm

Lipid-soluble substances

Water-soluble substances

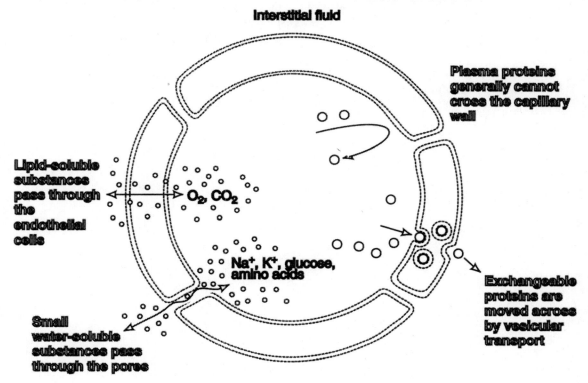

Interstitial fluid

Plasma proteins generally cannot cross the capillary wall

Lipid-soluble substances pass through the endothelial cells

O₂, CO₂

Na⁺, K⁺, glucose, amino acids

Exchangeable proteins are moved across by vesicular transport

Small water-soluble substances pass through the pores

Transport across capillary wall

10-18 Capillary bed.

Capillaries branch either directly from an arteriole or from a metarteriole, a thoroughfare channel between an arteriole and venule. Capillaries rejoin at either a venule or a metarteriole. Metarterioles are surrounded by smooth muscle cells, which also form precapillary sphincters that encircle capillaries as they arise from a metarteriole. (a) When the precapillary sphincters are relaxed, blood flows through the entire capillary bed. (b) When the precapillary sphincters are contracted, blood flows only through the metarteriole, bypassing the capillary bed.

Color

Smooth muscles

Arteriole

Venule

Directional arrows

Label

Capillaries

Metarteriole

Precapillary sphincter

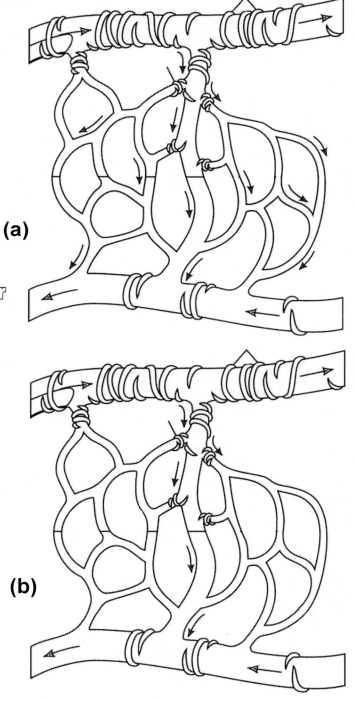

(a)

(b)

10-22 Lymphatic system.

(a) Lymph empties into the venous system near its entrance to the right atrium.
(b) Lymph flow averages 3 liters per day, whereas blood flow averages 7200 liters per day.

Color

Lymph vessel

Lymph node

Initial lymphatics

Arteries

Veins

Label

Blood Capillaries

Systemic circulation

Pulmonary circulation

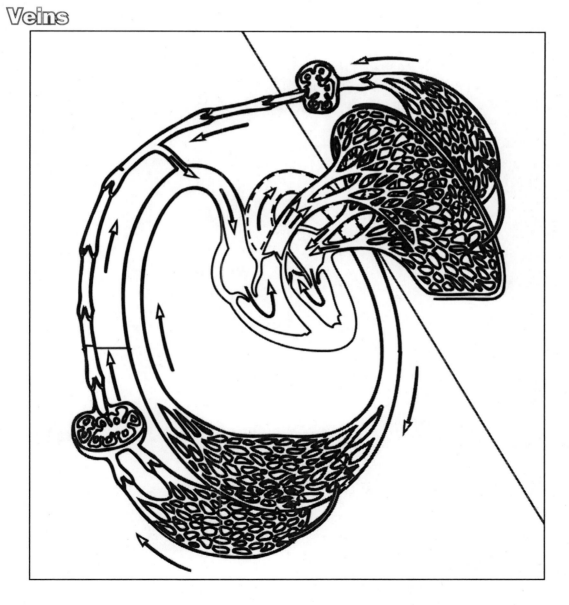

10-27 Function of venous valves.

(a) When a tube is squeezed in the middle, fluid is pushed in both directions. (b) Venous valves permit the flow of blood only toward the heart.

Color

Muscles

Vein walls

Valves

Interiors of veins

Blood flow arrows

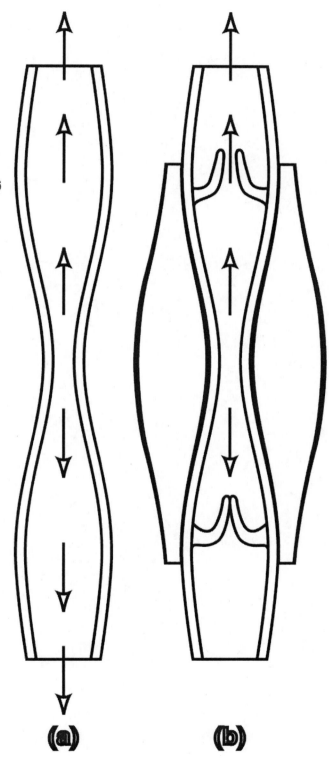

(a)   (b)

11-1 Hematocrit and types of blood cells.

The values given are for men. The average hematocrit for
a woman is 42%, with plasma occupying 58% of the blood volume.

Color

Plasma

Red blood cells

Buffy coat

Label

Platelets

White Blood cells

Hematocrit

List the percentage of each part of whole blood:

Plasma =_____%

Red blood cells =_____%

Buffy coat =_____%

11-2 Hemoglobin molecule.

A hemoglobin molecule consists of four highly folded polypeptide chains (the globin portion) and four iron-containing heme groups.

Heme groups

α Polypeptide chain 1

α Polypeptide chain 2

β Polypeptide chain 1

β Polypeptide chain 2

Label the atoms in the iron-containing heme group

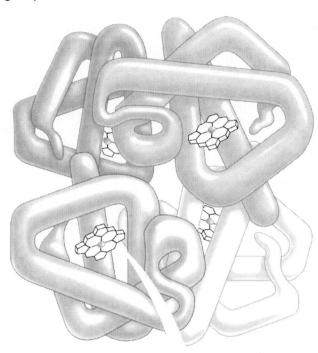

11-3 Major steps in erythrocyte production (erythropoiesis).

Erythrocytes are derived in the red bone marrow from pluripotent stem cells that give rise to all the types of blood cells. Myeloid stem cells are partially differentiated cells that give rise to erythrocytes and several other types of blood cells. Nucleated erythroblasts are committed to becoming mature erythrocytes. These cells extrude their nucleus and organelles, making more room for hemoglobin. Reticulocytes are immature red blood cells that contain organelle remnants. Mature erythrocytes are released into the abundant capillaries in the bone marrow.

Color and label

Pluripotent stem cell

Myeloid stem cell

Erythroblast

Nucleus and organelles

Reticulocyte

Erthrocyte

No nucleus or organelles

Remnants of organelles

11-4 Control of erythropoiesis.

Color

Kidney

Developing
   erythrocytes in red
   bone marrow

Erythrocytes

Reduced oxygen
   carrying capacity

Increased oxygen
   carrying capacity

Erythropoeitin

Blood vessels

Bone

Lymph vessel

Label and describe the 5 steps in this process:

1._____
_____
_____

2._____
_____
_____

3._____
_____
_____

4._____
_____
_____

5._____
_____
_____

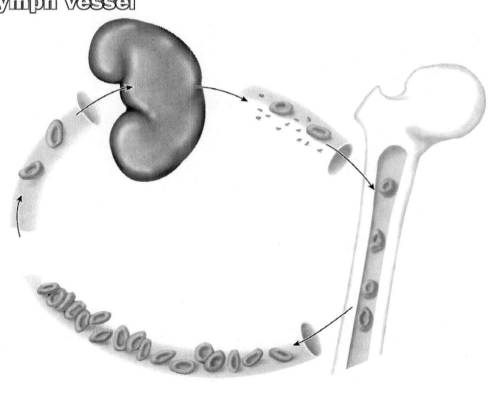

11-5 Hematocrit under various circumstances.

(a) Normal hematocrit. (b) The hematocrit is lower than normal in anemia, because of too few circulating erythrocytes; and (c) above normal in polycythemia, because of excess circulating erythrocytes. (d) The hematocrit can also be elevated in dehydration when the normal number of circulating erythrocytes is concentrated in a reduced plasma volume.

Color

Plasma

Erythrocytes

Label type of Hematocrit and the percentage

Normal

Anemic

Polycythemia

Dehydration

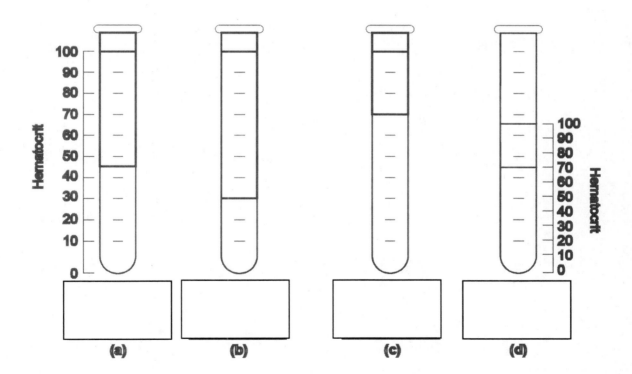

11-7 Transfusion reaction.

A transfusion reaction resulting from type B blood being transfused into a recipient with type A blood.

Color

Donor with type B blood
Antigen B
Antibody to type A blood
Antibody to Type B blood
Recipient with type A blood
Antigen A

Label

Red blood cells from
    donor agglutinate
Red blood cells usually rupture
Clumping blocks blood
    flow in capillaries
Oxygen and nutrient
    flow to cells and
    tissues is reduced
Hemoglobin
    precipitates in
    kidney, interfering
    with kidney function

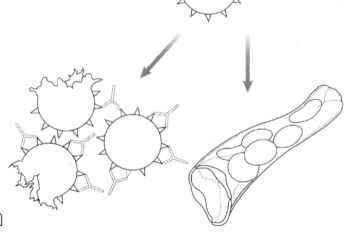

11-11 Formation of a platelet plug.

Platelets aggregate at a vessel defect through a positive-feedback mechanism involving the release of adenosine diphosphate (ADP) and thromboxane A2 from platelets, which stick to exposed collagen at the site of the injury. Platelets are prevented from aggregating at the adjacent normal vessel lining by the release of prostacyclin and nitric oxide from the undamaged endothelial cells.

Color

Platelet

Vessel lumen

Interstitial fluid

Endothelium

Smooth muscle

Outerconnective tissue layer

Subendothelial connective tissue

Label

Exposed collagen at site of vessel injury

Vessel wall

Normal endothelium

Thromboxane A2

Adenosine diphosphate

Prostacyclin and nitric oxide

Aggregating Collagen platelet plug

Label and describe the 5 steps in this process

1. _____

2. _____

3. _____

4. _____

5. _____

12-2 Steps producing inflammation.

Chemotaxins released at the site of damage attract phagocytes to the scene. Note the leukocytes emigrating from the blood into the tissues by assuming amoeba-like behavior and squeezing through the capillary pores, a process known as diapedesis. Mast cells secrete vessel-dilating, pore-widening histamine. Macrophages secrete cytokines that exert multiple local and systemic effects.

Color

Macrophage
Monocyte
Cytokines
Chemotaxins
Histamine

Mast cells
Bacteria at injury site
Neutrophils
Capillary
Skin cells

Label

Process of diapedesis
Neutrophils sticking to wall
Endothelial cell of capillary

Label the 5 steps in this process and describe them on the back of this page.

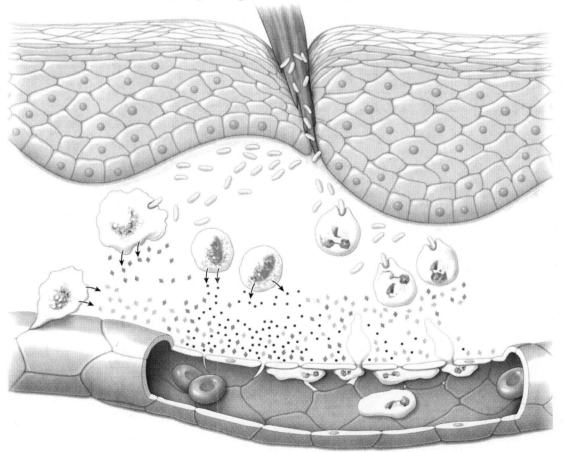

12-4 Mechanism of opsonin action.

One of the activated complement molecules, C3b, links a foreign cell, such as a bacterium, and a phagocytic cell by nonspecifically binding with the foreign cell and specifically binding with a receptor on the phagocyte. This link ensures that the foreign victim does not escape before it can be engulfed by the phagocyte.

Color

Bacterium membrane

Bacterium interior

Activated complement

Receptor

Phagocyte

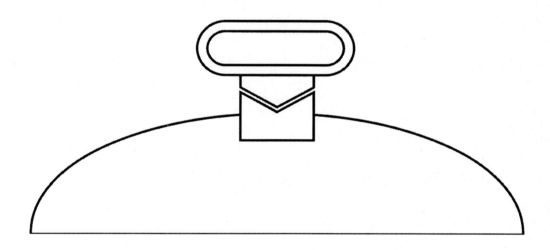

Structures are not drawn to scale.

12-10 Antibody structure.

An antibody is Y-shaped. It is able to bind only with the specific antigen that "fits" its antigen-binding sites (Fab) on the arm tips. The tail region (Fc) binds with particular mediators of antibody-induced activities.

Color

Triangle antigen
Rectangle antigen
U shaped antigen
Half oval antigen
Stepped antigen
M shaped antigen
Variable region
Constant region

Label

Specific antigen-
binding sites
Light chain
Heavy Chain
Antibody
Disulfide

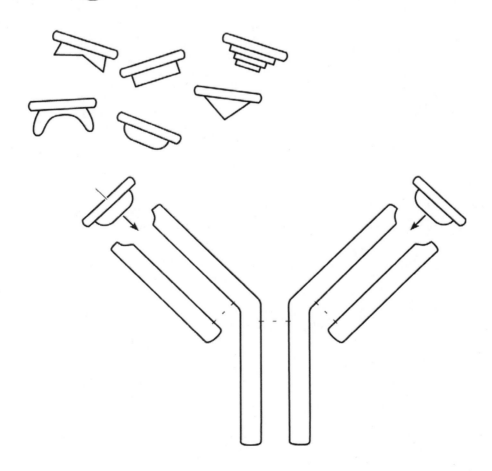

12-11 How antibodies help eliminate invading microbes.

Antibodies physically hinder antigens through (1) neutralization or (2) agglutination and precipitation. Antibodies amplify innate immune responses by (1) activating the complement system, (2) enhancing phagocytosis by acting as opsonins, and (3) stimulating killer cells.

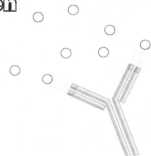

Color

Invading bacterium

Toxin

Antibody produced against toxin antigen

Antibody

Invading bacterium

Membrane attack complex

Inactive C1 complement molecule

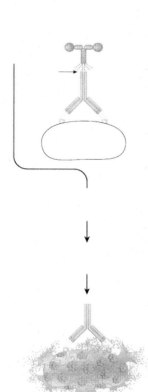

Label

Lysis of cell

Binds with

Activated by binding with
   antigen-attached antibody

Formation of C5–C9, the
   membrane attack complex

12-11 (continued)

Color

Foreign cells
Antigen
Antibodies
Invading bacterium
Phagocyte
Natural killer cell

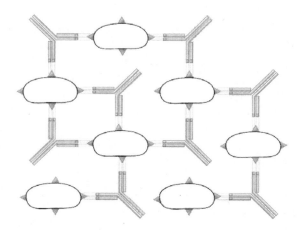

Label

Lysis induced by
killer cell

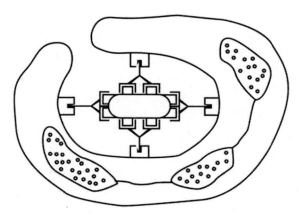

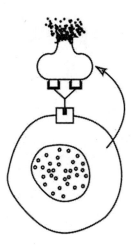

12-12 Clonal selection theory.

The B-cell clone specific to the antigen proliferates and differentiates into plasma cells and memory cells. Plasma cells secrete antibodies that bind with free antigen not attached to B cells. Memory cells are primed and ready for subsequent exposure to the same antigen.

Color

Antigen

BCR

Plasma cells

Memory B cells

Antibodies

Rough ER

B cell specific to antigen

Nucleus

Antibodies

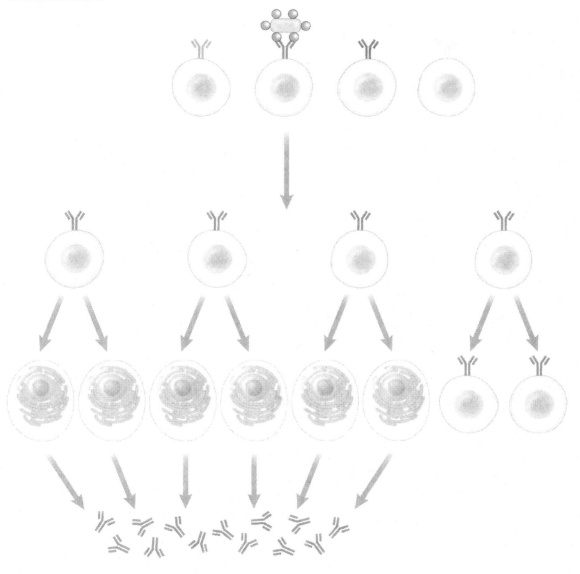

© 2016 Cengage Learning. All Rights Reserved. May not be scanned, copied or duplicated, or posted to a publicly accessible website, in whole or in part.

12-14 A cytotoxic T cell lysing a virus-invaded cell.

Color

Virus

Viral antigenic protein coat

Host cell

Foreign viral antigen

MHC self-antigen

Virus invaded host cell

Cytotoxic T cell

T-cell receptor

MHC self-antigen and foreign
   antigen complex

Virus invaded host cell

Describe the 4 steps in this process:

1._____
_____
_____

2._____
_____
_____

3._____
_____
_____

4._____
_____
_____

12-15 Mechanism of killing by killer cells.

Note the similarity of the perforin- formed pores in a target cell to the membrane attack complex formed by complement molecules (see Figure 12-6b, p. 425). Source: Adapted from the illustration by Dana Burns-Pizer in "How Killer Cells Kill," by John Ding-E Young and Zanvil A. Cohn in Scientific America, 1988.)

Color

Granule containing
   perforin molecules

Nucleus

Killer cell

Target cell

Cell membrane –
   heads

Cell membrane – tails

Perforin

Granule

Virus

Label and describe the 7 steps in this process:

1._____
_____

2._____
_____

3._____
_____

4._____
_____

5._____
_____

6._____
_____

7._____
_____

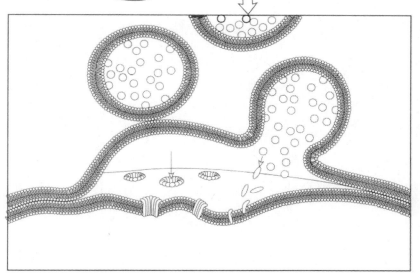

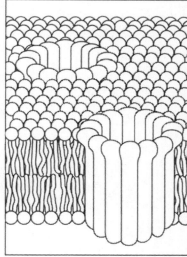

12-18 Generation of an antigen-presenting cell when a dendritic cell engulfs a bacterium.

Color

Antigen

Lysosome

Compartment for peptide loading

Bacterium

Dendritic cell

MHC protein

Antigen-MHC complex on cell surface

Label and describe the 5 steps in this process:

1._____
_____
_____

2._____
_____
_____

3._____
_____
_____

4._____
_____
_____

5._____
_____
_____

237

12-19 Distinctions between class I and class II major histocompatibility complex (MHC) glycoproteins.

Specific binding requirements for the two types of T cells ensure that these cells bind only with the target cells with which they can interact. Cytotoxic (CD81) T cells can recognize and bind with foreign antigen only when the antigen is in association with class I MHC glycoproteins, which are found on the surface of all body cells. This requirement is met when a virus invades a body cell, whereupon the cell is destroyed by the cytotoxic T cells. Helper T (CD41) cells, which are activated by and/or enhance the activities of dendritic cells, macrophages, and B cells, can recognize and bind with foreign antigen only when it is in association with class II MHC glycoproteins, which are found only on the surface of these other immune cells. The CD81 or CD41 T-cell's coreceptor CD8 or CD4 links these cells to the target cell's class I or class II MHC molecules, respectively.

Color

Bacterium

Antigens

T-cell receptor (TCR)

CD4 coreceptor

Helper T cells

Class II MHC protein

Dendritic cell

Helper (CD4+) T cell

Interleukins

Cytokines

Label and describe the 4 steps associated with each process

1._____

_____

2._____

_____

3._____

_____

4._____

_____

1._____

_____

2._____

_____

3._____

_____

4._____

_____

239

12-23 Role of IgE antibodies and mast cells in immediate hypersensitivity.

B-cell clones are converted into plasma cells, which secrete IgE antibodies on contact with the allergen for which they are specific. All IgE antibodies, regardless of their antigen specificity, bind to mast cells or basophils. When an allergen combines with the IgE receptor specific for it on the surface of a mast cell, the mast cell releases histamine and other chemicals by exocytosis. These chemicals elicit the allergic response.

Specific B-cell clones
IgE antibodies
Mitochondrion
IgE tail receptor
Histamine granules
B cell
Allergens
Activated Plasma cells
Mast cell
Histamine release

Label and describe the 5 steps in this process:

1._____
_____
_____

2._____
_____

3._____
_____
_____

4._____
_____
_____

5._____
_____
_____

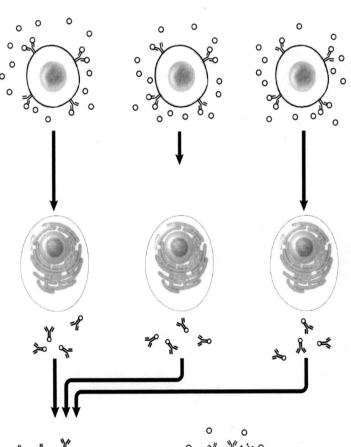

12-24 Anatomy of the skin

The skin consists of two layers, a keratinized outer epidermis and a richly vascularized inner connective tissue dermis. Special infoldings of the epidermis form the sweat glands, sebaceous glands, and hair follicles. The epidermis contains four types of cells: keratinocytes, melanocytes, Langerhans cells, and Granstein cells. The skin is anchored to underlying muscle or bone by the hypodermis, a loose, fat-containing layer of connective tissue.

Color
Adipose cells
Nerve fiber
Pressure receptor
T lymphocyte
Granstein cell
Hair shaft
Langerhans cell
Melanocyte

Label
Dermis
Epidermis
Hypodermis
Hair follicle
Living layer
Sebaceous gland

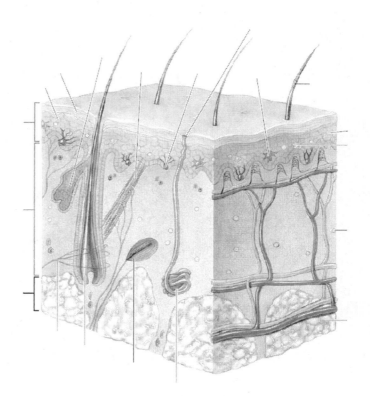

Keratinized layer
Keratinocyte

Smooth muscle
Sweat gland

What layers of skin are directly supplied by the blood? What are the implications of this vascular pattern?

_____

_____

_____

_____

13-1 External and internal respiration.

External respiration encompasses the steps involved in the exchange of O2 and CO2 between the external environment and tissue cells (steps 1 through 4). Internal respiration encompasses the intracellular metabolic reactions involving the use of O2 to derive energy (ATP) from food, producing CO2 as a by-product.

Color

Lungs

Tissue cells

O_2 rich blood

O_2 poor blood

Heart

Describe the 4 steps involved in this process:

1. _____

_____

_____

_____

2. _____

_____

_____

_____

3. _____

_____

_____

_____

4. _____

_____

_____

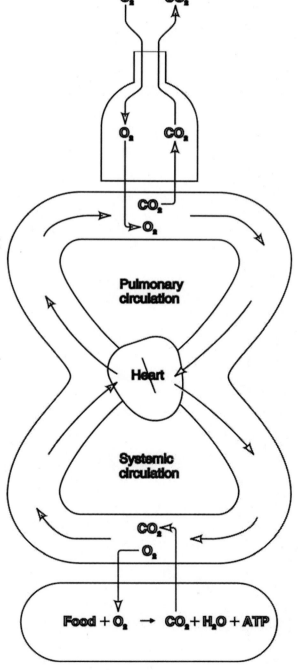

Pulmonary circulation

Heart

Systemic circulation

$Food + O_2 \rightarrow CO_2 + H_2O + ATP$

245

13-4 Alveolus and associated pulmonary capillaries.

(a) A single layer of flattened Type I alveolar cells forms the alveolar walls. Type II alveolar cells embedded within the alveolar wall secrete pulmonary surfactant. Wandering alveolar macrophages are found within the alveolar lumen. The size of the cells and respiratory membrane is exaggerated compared to the size of the alveolar and pulmonary capillary lumens. The diameter of an alveolus is actually about 600 times larger (300 mm) than the intervening space between air and blood (0.5 mm).

Color

Alveolar fluid lining
    with pulmonary
      surfactant
Alveolar macrophage
Erythrocyte
Monocyte
Pulmonary capillary

Label

0.5 μm barrier
    separating air
    and blood
Elastin fiber

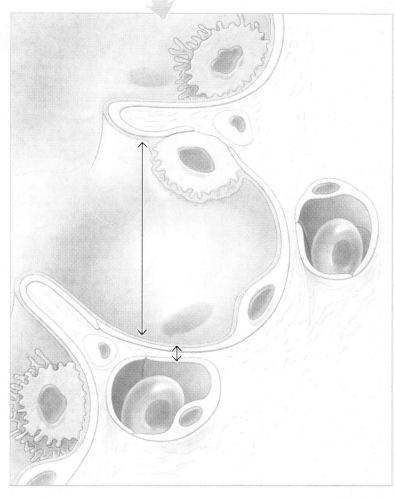

13-8 Pneumothorax.

(a) *Traumatic pneumothorax.* A puncture in the chest wall permits air from the atmosphere to flow down its pressure gradient and enter the pleural cavity, abolishing the transmural pressure gradient. (b) *Collapsed lung.* When the transmural pressure gradient is abolished, the lung collapses to its unstretched size, and the chest wall springs outward. (c) *Spontaneous pneumothorax.* A hole in the lung wall permits air to move down its pressure gradient and enter the pleural cavity from the lungs, abolishing the transmural pressure gradient. As with traumatic pneumothorax, the lung collapses to its unstretched size.

Color

Pleural sacs

Trachea

Right lung

Left lung

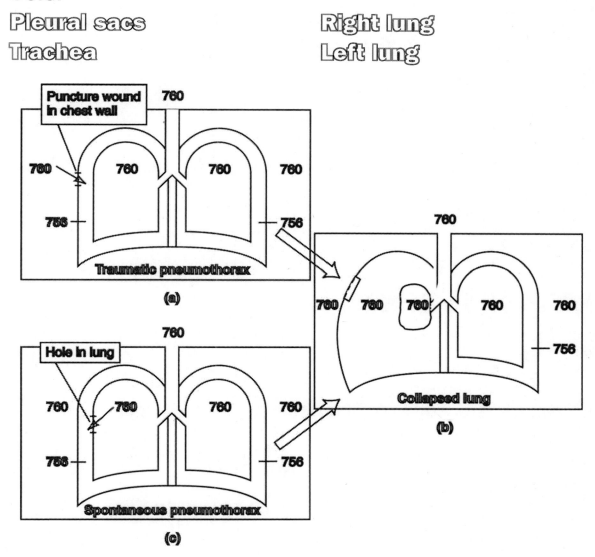

Numbers are mm Hg pressure.

13-9 Boyle's law.

Each container has the same number of gas molecules. Given the random motion of gas molecules, the likelihood of a gas molecule striking the interior wall of the container and exerting pressure varies inversely with the volume of the container at any constant temperature. The gas in container B exerts more pressure than the same gas in larger container C but less pressure than the same gas in smaller container A. This relationship is stated as Boyle's law: $P1V1 \_ P2V2$. As the volume of a gas increases, the pressure of the gas decreases proportionately; conversely, the pressure increases proportionately as the volume decreases.

Color

Gas molecules
Piston

Space between
molecules
Pressure gauge

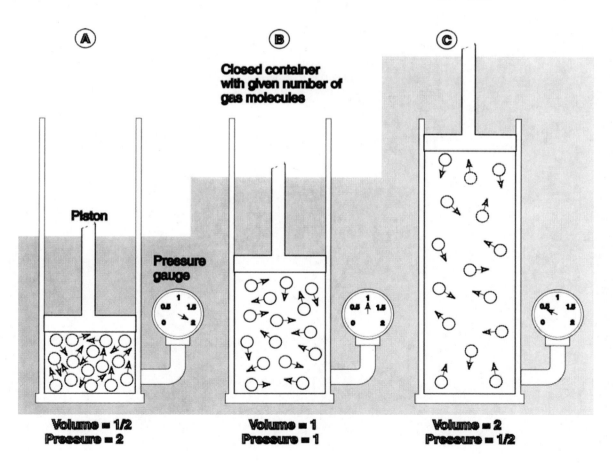

Ⓐ

Ⓑ Closed container with given number of gas molecules

Ⓒ

Piston

Pressure gauge

Volume = 1/2
Pressure = 2

Volume = 1
Pressure = 1

Volume = 2
Pressure = 1/2

13-14 Alveolar interdependence.

(a) When an alveolus *(in pink)* in a group of interconnected alveoli starts to collapse, the surrounding alveoli are stretched by the collapsing alveolus. (b) As the neighboring alveoli recoil in resistance to being stretched, they pull outward on the collapsing alveolus. This expanding force pulls the collapsing alveolus open.

Color

Alveoli

Directional arrows

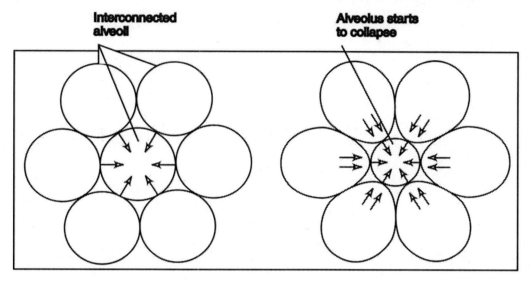

Interconnected alveoli

Alveolus starts to collapse

(a)

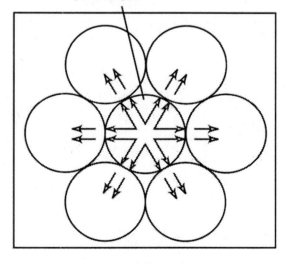

Collapsing alveolus pulled open

(b)

13-25 Hemoglobin facilitating a large net transfer of O_2 by acting as a storage depot to keep PO_2 low.

(a) In the hypothetical situation in which no hemoglobin is present in the blood, the alveolar PO_2 and the pulmonary capillary blood PO_2 are at equilibrium. (b) Hemoglobin has been added to the pulmonary capillary blood. As the Hb starts to bind with O_2, it removes O_2 from solution. Because only dissolved O_2 contributes to blood PO_2, the blood PO_2 falls below that of the alveoli, even though the same number of O_2 molecules are present in the blood as in part (a). By "soaking up" some of the dissolved O_2, Hb favors the net diffusion of more O_2 down its partial pressure gradient from the alveoli to the blood. (c) Hemoglobin is fully saturated with O_2, and the alveolar and blood PO_2 are at equilibrium again. The blood PO_2 resulting from dissolved O_2 is equal to the alveolar PO_2, despite the fact that the total O_2 content in the blood is much greater than in part (a) when blood PO_2 was equal to alveolar PO_2 in the absence of Hb.

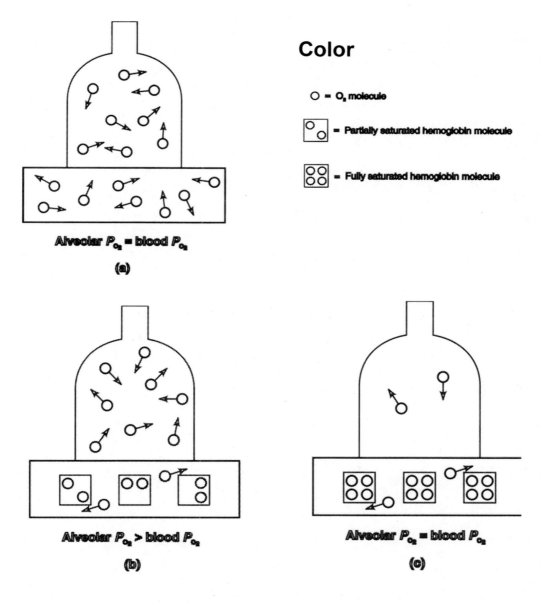

Color

O = O_2 molecule

= Partially saturated hemoglobin molecule

= Fully saturated hemoglobin molecule

Alveolar P_{O_2} = blood P_{O_2}

(a)

Alveolar P_{O_2} > blood P_{O_2}

(b)

Alveolar P_{O_2} = blood P_{O_2}

(c)

13-29 Respiratory control centers in the brain stem.

Color

Pneumotaxic center
Apneustic center
Pre-Botzinger complex
Dorsal respiratory group
Ventral respiratory group
Medulla
Pons

Label

Medullary respiratory center
Pons respiratory centers
Respiratory control centers in brain stem

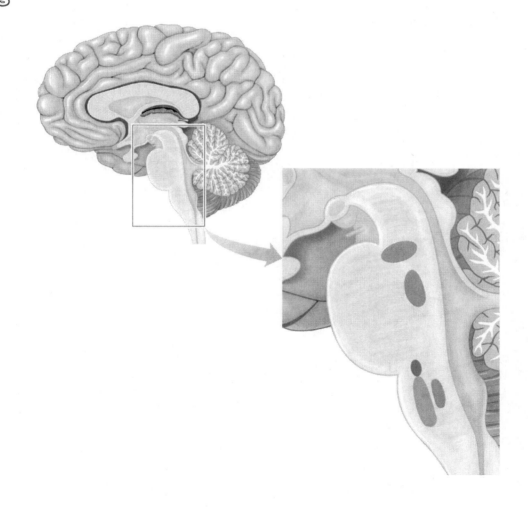

13-30 Location of peripheral chemoreceptors

The carotid bodies are located in the carotid sinus, and the aortic bodies are located in the aortic arch.

The arrows below indicate flow to what particular region?

_____

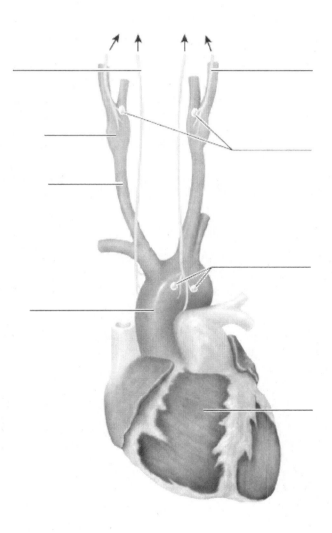

Color

Aortic arch

Heart

Sensory nerve fiber

Label

Aortic bodies

Carotid artery

Carotid bodies

Carotid sinus

14-1 The urinary system.

(a) The pair of kidneys forms the urine, which the ureters carry to the urinary bladder. Urine is stored in the bladder and periodically emptied to the exterior through the urethra. (b) The kidney consists of an outer, granular appearing renal cortex and an inner, striated-appearing renal medulla. The renal pelvis at the medial inner core of the kidney collects urine after it is formed.

Color

Renal vein

Inferior vena cava

Urinary bladder

Urethra

Renal artery

Kidney

Aorta

Ureter

Renal cortex

Renal medulla

Renal pelvis

Renal pyramid

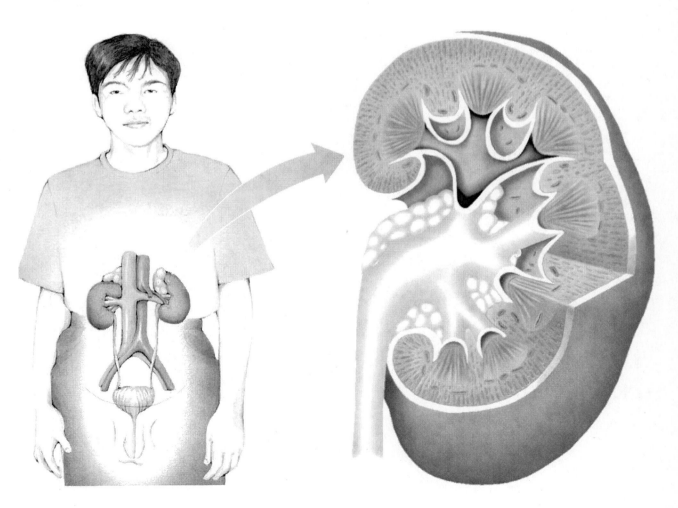

14-3 A nephron.

A schematic representation of a cortical nephron, the most abundant type of nephron in humans.

Color

Afferent arteriole
Glomerulus
Efferent arteriole
Peritubular capillaries
Juxtaglomerular apparatus
Bowman's capsule
Proximal tubule
Loop of Henle
Distal tubule
collecting duct

Cortex
Medulla

To renal pelvis

14-5 Comparison of juxtamedullary and cortical nephrons.

The glomeruli of cortical nephrons lie in the outer cortex, whereas the glomeruli of juxtamedullary nephrons lie in the inner part of the cortex next to the medulla. The loops of Henle of cortical nephrons dip only slightly into the medulla, but the juxtamedullary nephrons have long loops of Henle that plunge deep into the medulla. The juxtamedullary nephrons' peritubular capillaries form hairpin loops known as *vasa recta.*

Color

Proximal tube

Distal tube

Loop of Henle

Glomerulus

Bowman's capsule

Collecting duct

Vasa reta

Medulla

Cortex

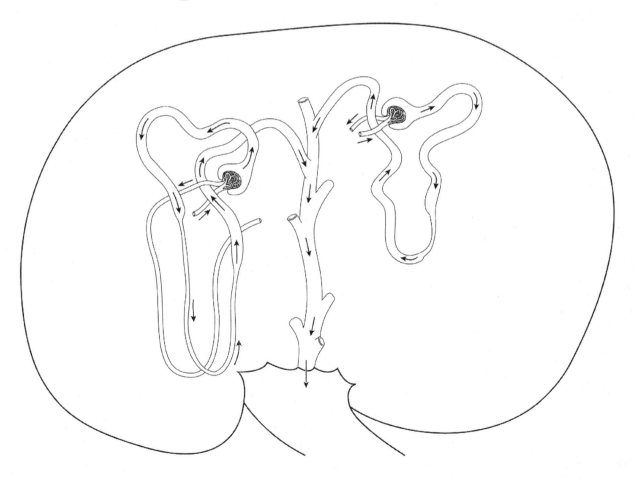

14-6 Basic renal processes.

Anything filtered or secreted but not reabsorbed is excreted in the urine and lost from the body. Anything filtered and subsequently reabsorbed, or not filtered at all, enters the venous blood and is saved for the body.

Color

Glomerulus

Bowman's capsule

Afferent arteriole

Efferent arteriole

Kidney tubule

Pertubular capillary

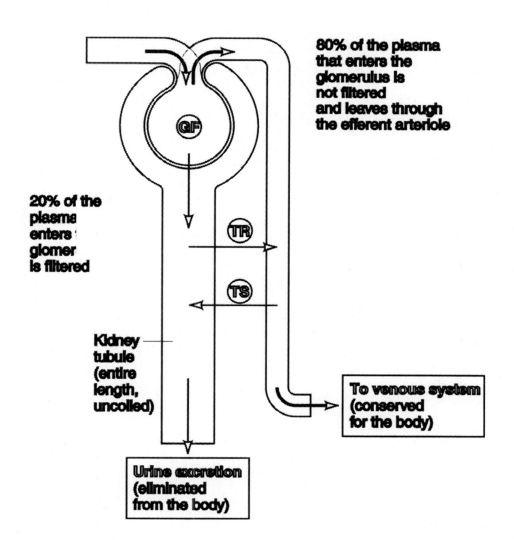

80% of the plasma that enters the glomerulus is not filtered and leaves through the efferent arteriole

GF

TR

TS

20% of the plasma enters glomer is filtered

Kidney tubule (entire length, uncoiled)

To venous system (conserved for the body)

Urine excretion (eliminated from the body)

14-7 Layers of the glomerular membrane.

Color

Afferent arteriole

Efferent arteriole

Glomerulus

Bowman's capsule

Lumen of Bowman's capsule

Outer layer of Bowman's capsule

Proximal convoluted tubule

Inner layer of Bowman's capsule

Basement membrane

Lumen of glomerular capillary

Filtration slit

Podocyte foot process

Capillary pore

Endothelial cell

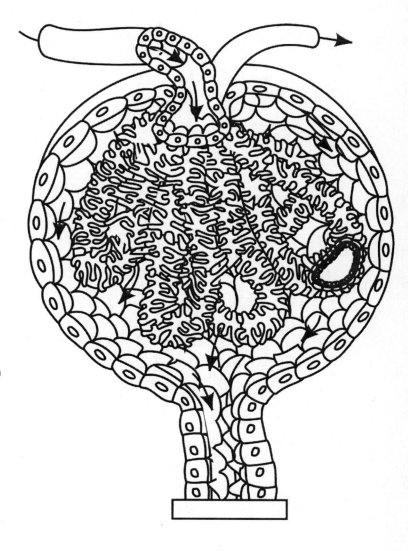

269

14-11 The juxtaglomerular apparatus.

The juxtaglomerular apparatus consists of specialized vascular cells (the granular cells) and specialized tubular cells (the macula densa) at a point where the distal tubule passes through the fork formed by the afferent and efferent arterioles of the same nephron.

Color

Afferemt arteriole

Efferent arteriole

Glomerular capillaries

Bowman's capsule

Lumen of Bowman's capsule

Distal tubule

Juxtaglomerular apparatus

Macula densa

Endothelial cell

Podocyte

Cranular cells

Endothelial cell

Smooth muscle cell

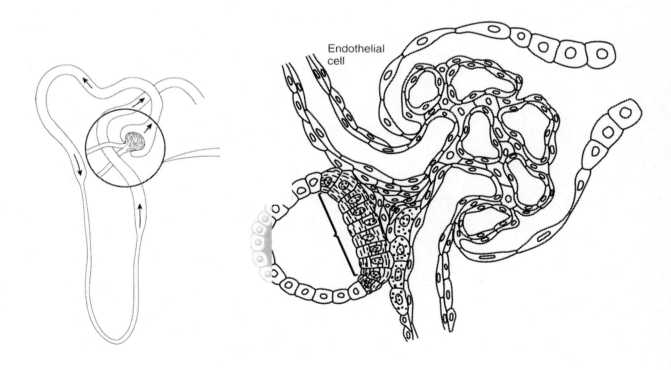

Endothelial cell

14-14 Steps of transepithelial transport.

Color
Plasma
Capillary wall
Lateral space
Filtrate
Tubular lumen

Interstitial fluid

Label
Basolateral membrane
Tight junction

List the 5 barriers a substance must transverse to be reabsorbed

1. _____

2. _____

3. _____

4. _____

5. _____

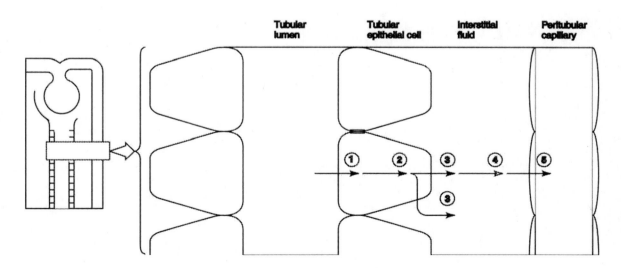

14-23 Plasma clearance for substances handled in different ways by the kidneys.

Color
Peritubular capillary
Glomerulus
Tubule
Round substance
Square substance
Triangular substance
Plus sign substance

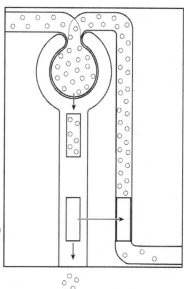
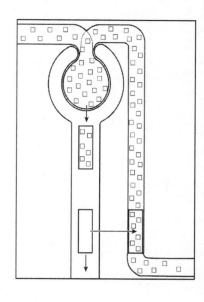
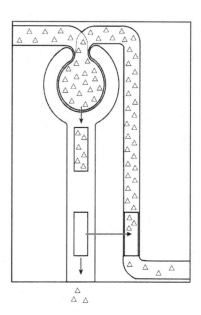
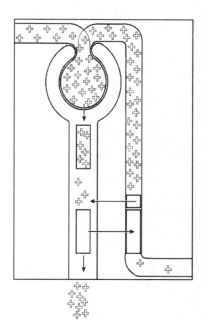

15-2 Ionic composition of the major body-fluid compartments.

Color

Potassium

Sodium

Chloride

Other

Protein anions

Bicarbonate

Phosphate

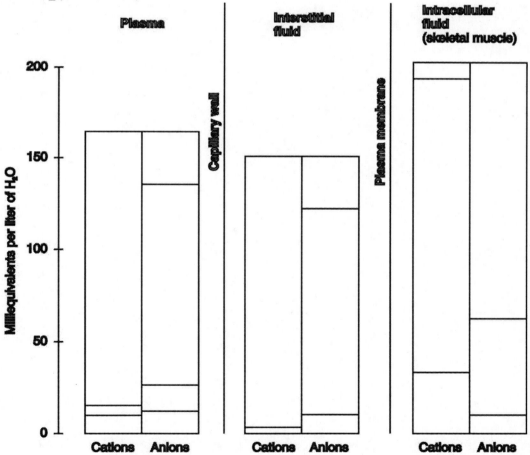

15-8 Action of chemical buffers.

(a) Addition of HCl to an unbuffered solution. All the added hydrogen ions (H+) remain free and contribute to the acidity of the solution. (b) Addition of HCl to a buffered solution. Bicarbonate ions (HCO3-), the basic member of the buffer pair, bind with some of the added H+ and remove them from solution so that they do not contribute to its acidity.

Color

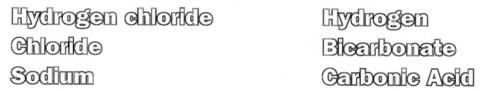

Hydrogen chloride
Chloride
Sodium

Hydrogen
Bicarbonate
Carbonic Acid

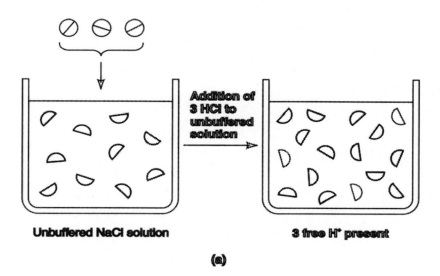

Unbuffered NaCl solution

Addition of 3 HCl to unbuffered solution

3 free H⁺ present

(a)

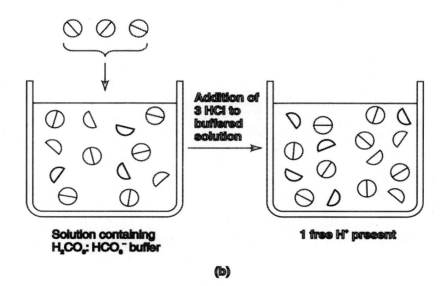

Solution containing H₂CO₃: HCO₃⁻ buffer

Addition of 3 HCl to buffered solution

1 free H⁺ present

(b)

279

16-2 Layers of the digestive tract wall.

The digestive tract wall consists of four major layers: from the innermost out, they are the mucosa, submucosa, muscularis externa, and serosa.

Color

Body wall
Peritoneum
Mesentery
Serosa
Lumen
Muscularis mucosa
Lamina propria

Label

Submucosal plexus
Myenteric plexus

Mucous membrane
Outer longitudinal muscle
Inner circular muscle
Submucosa
Duct of large accessory
    digestive gland emptying
    into digestive tract lumen

Mucosa
Muscularis externa

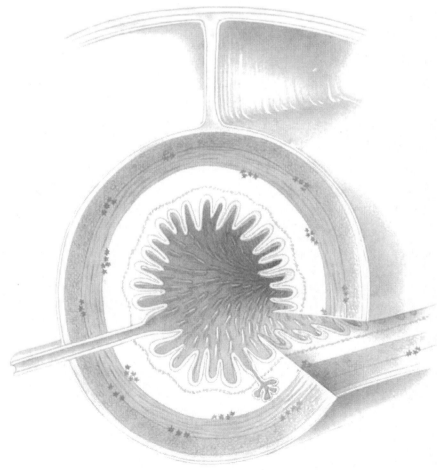

16-6 Anatomy of the stomach

The stomach is divided into three sections based on structural and functional distinctions – the fundus, body, and antrum. The mucosal lining of the stomach is divided into the oxyntic mucosa and the pyloric gland area based on differences in glandular secretion.

Color

Duodenum

Esophagus

Oxyntic mucosa

Phyloric gland area

Smooth muscle

Label

Antrum

Body

Fundus

Gastroesophageal sphincter

Pyloric sphincter

Stomach folds

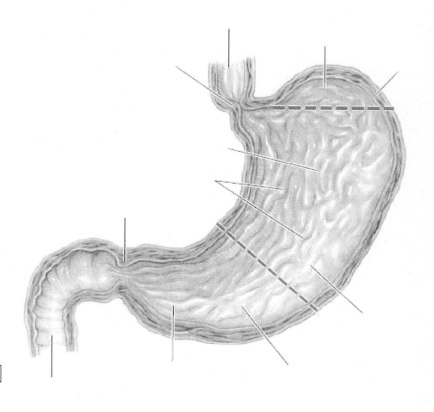

Name three main functions of the stomach.

_____

_____

_____

_____

_____

16-7 Gastric emptying and mixing as a result of antral peristaltic contractions.

Color

Stomach
Esophagus
Gastroesophageal
sphincter
Pyloric sphincter
Duodenum

Label

Movement of chime
Peristaltic contraction
Direction of movement

List the 6 steps of this process

1. _____
_____
_____

2. _____
_____
_____

3. _____
_____
_____

4. _____
_____
_____

5. _____
_____
_____

6. _____
_____
_____

movement
of chyme

of peristaltic
contraction

Gastric mixing

16-10 Exocrine and endocrine portions of the pancreas.

The exocrine pancreas secretes into the duodenal lumen a digestive juice composed of digestive enzymes secreted by the acinar cells and an aqueous NaHCO3 solution secreted by the duct cells. The endocrine pancreas secretes the hormones insulin and glucagon into the blood.

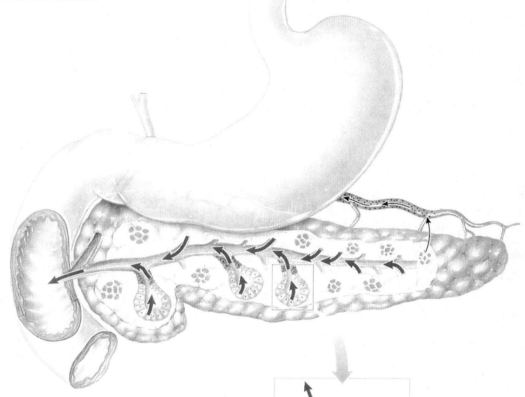

Color

Stomach

Bile duct from liver

Duodenum

Duct cells

Acinar cells

Endocrine portion of pancreas

Blood

Label

Exocrine portion of pancreas

Hormones

Zymogen granule

16-13 Schematic representation of liver blood flow.

Color

Digestive capillaries
Digestive tract
Liver
Liver sinusoids
Inferior vena cava
Heart
Hepatic vein
Hepatic artery
Hepatic portal vein
Aorta
Arteries to digestive tract

List the 2 steps of this process

1a._____

_____

1b._____

_____

_____

2. _____

_____

_____

16-14 Anatomy of the liver.

Color

Branch of hepatic portal vein
Bile duct
Branch of hepatic artery
Sinusoids
Bile canaliculi
Central vein
Cord of hepatocytes
Hepatic portal vein
Hepatic artery
Kupffer cell
Connective tissue

Label

To hepatic duct
Hepatic plate

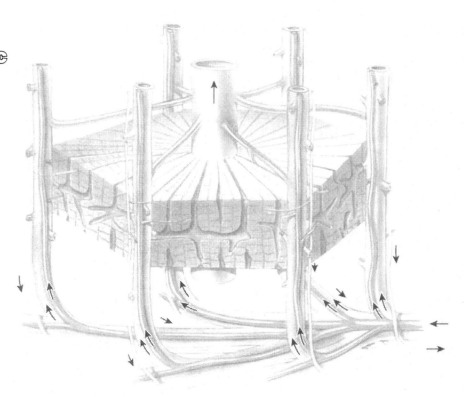

16-15 Enterohepatic circulation of bile salts.

The majority of bile salts are recycled between the liver and small intestine through the enterohepatic circulation *(blue arrows)*. After participating in fat digestion and absorption, most bile salts are reabsorbed by active transport in the terminal ileum and returned through the hepatic portal vein to the liver, which resecretes them in the bile.

Color

Portal circulation

Gallblader

Sphincter of Oddi

Colon

Terminal ileum

Bile salts

Cholesteral

Common bile duct

Duodenum

Liver

Directional arrows

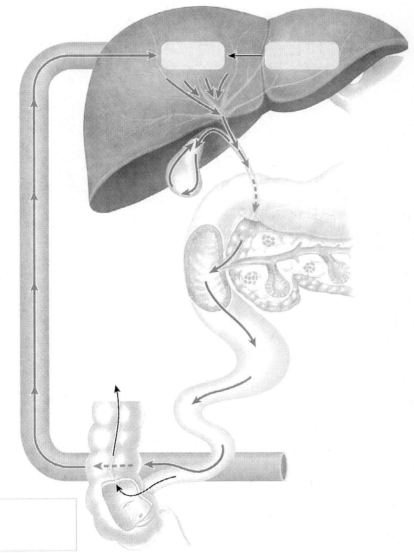

16-16 Schematic structure and function of bile salts.

(a) Schematic representation of the structure of bile salts and their adsorption on the surface of a small fat droplet. A bile salt consists of a lipid-soluble part that dissolves in the fat droplet and a negatively charged, water-soluble part that projects from the surface of the droplet. (b) Formation of a lipid emulsion through the action of bile salts. When a large fat droplet is broken up into smaller fat droplets by intestinal contractions, bile salts adsorb on the surface of the small droplets, creating shells of negatively charged, water-soluble bile salt components that cause the fat droplets to repel one another. This action holds the fat droplets apart and prevents them from recoalescing, increasing the surface area of exposed fat available for digestion by pancreatic lipase. Such emulsified fat droplets are about 1 mm in diameter.

Color

Large fat droplet

Lipid emulsion

Lipid soluble head

Water soluble tail

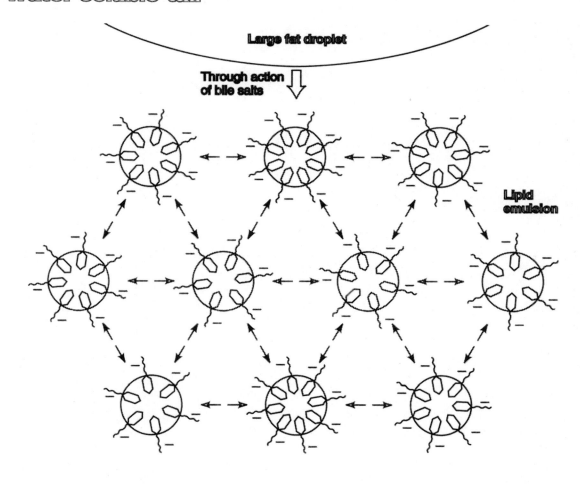

16-17 A micelle.

Bile constituents (bile salts, lecithin, and cholesterol) aggregate to form micelles that consist of a hydrophilic (water-soluble) shell and a hydrophobic (lipid-soluble) core. Because the outer shell of a micelle is water soluble, the products of fat digestion, which are not water soluble, can be carried through the watery luminal contents to the absorptive surface of the small intestine by dissolving in the micelle's lipid-soluble core. This figure is not drawn to scale compared to the lipid emulsion droplets in I Figure 16-17b. An emulsified fat droplet ranges in diameter from 200 to 5000 nm (average 1000 nm) compared to a micelle, which is 3 to 10 nm in diameter.

Color

Hydrophobic core

Hydrophilic shell

Color and label in Key

Water-soluble portion

Lipid-soluble portion

Cholesterol

Bile salt

All lipid-soluble

Lecithin

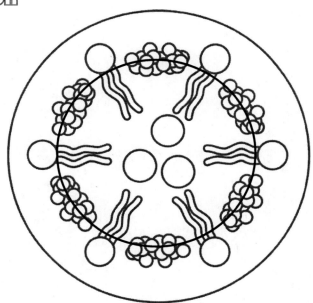

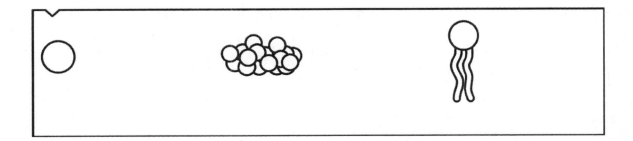

16-19 Small-intestine absorptive surface.

(a) Gross structure of the small intestine. (b) The circular folds of the small-intestine mucosa collectively increase the absorptive surface area threefold. (c) Microscopic fingerlike projections known as a villi collectively increase the surface area another 10-fold. (d) Each epithelial cell on a villus has microvilli on its luminal border; the microvilli increase the surface area another 20-fold. Together, these surface modifications increase the small intestine's absorptive surface area 600-fold.

Color

Small intestine

Circular fold

Villus

Epithelial cell

Lymphatic vessel

Venule

Arteriole

Crypt of Lieberkuhn

Central lacteal

Mucous cell

Epithelial cell

Capillaries

Nucleus

Endoplasmic Reticulum

Mitochondrion

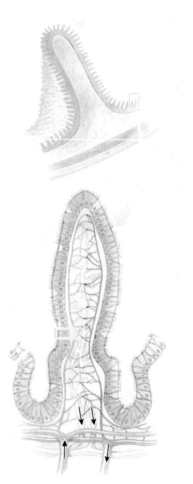

16-22 Carbohydrate digestion and absorption.

Color

Digestive-tract lumen

SGLT

Capillary

Interstitial fluid

Epithelial cell of villus

Glucose

Fructose

Galactose

GLUT-5

GLUT-2

Potassium

ATP

List the 6 steps of this process

1. _____

_____

2. _____

_____

3. _____

_____

4. _____

_____

5. _____

_____

6.

_____

_____

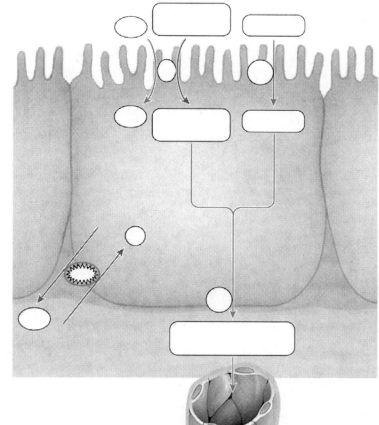

16-23 Protein digestion and absorption.

Color

Digestive-tract lumen   Intracellular peptidases

Capillary   Pepsin

Small peptides   Pancreatic enzymes

Amino acids   Epithelial cell of villus

Interstitial fluid   Hydrogen ions

Potassium   Aminopeptidases

ATP

List the 7 steps of this process

1. _____

_____

_____

2. _____

_____

_____

3. _____

_____

_____

4. _____

_____

_____

5. _____

_____

_____

6. _____

_____

_____

7. _____

_____

_____

16-24 Fat digestion and absorption.

Because fat is not soluble in water, it must undergo a series of transformations in order to be digested and absorbed.

Color

Micelle

Basement membrane

Lumen

Central lacteal

Microvillus

Fatty acids

Epithelial cell of villus

Dietary fat as large
triglyceride droplet

Lipid emulsion

Micelles

Chylomicrons

Passive absorption

Bile salts

Interstitial fluid

Blood capillary

Micellular diffusion

Monoglycerides

Free fatty acids

Triglycerides

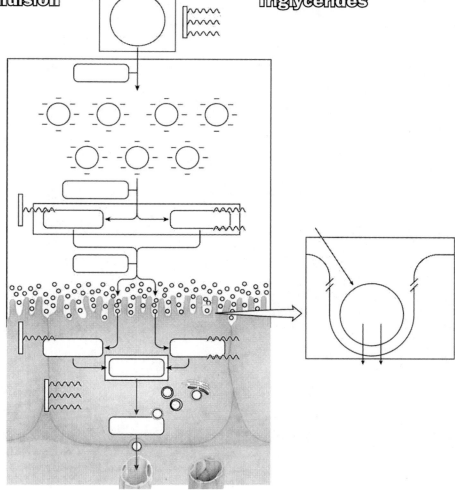

18-1 The endocrine system.

Color

Pineal

Hypothalamus

Pituitary

Parathyroid

Thyroid

Thymus

Heart

Stomach

Adrenal gland

Pancreas

Duodenum

Kidney

Adipose tissue

Skin

Ovaries in female

Placenta in pregnant
female

Testes in male

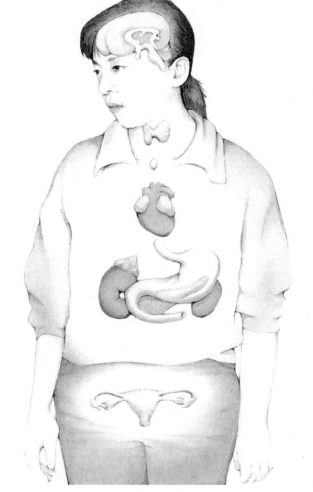

18-3 Anatomy of the pituitary gland.

(a) Relation of the pituitary gland to the hypothalamus and to the rest of the brain. (b) Enlargement of the pituitary gland and its connection to the hypothalamus.

Color

Hypothalamus

Bone

Anterior pituitary

Posterior pituitary

Optic chiasm

Hypothalamus

Connecting stalk

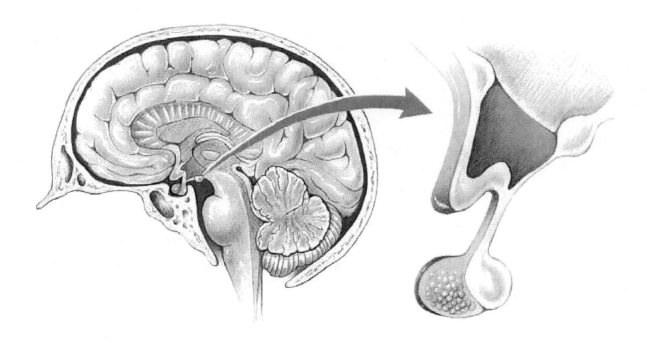

18-4 Relationship of the hypothalamus and posterior pituitary.

Color

Hypothalamus

Posterior pituitary

Systemic arterial inflow

Systemic venous
outflow

Vasopressin

Oxytocin

Paraventricular nucleus

Neurosecretory
neurons

Supraoptic nucleus

Hypothalamic posterior
pituitary stalk

Anterior pituitary

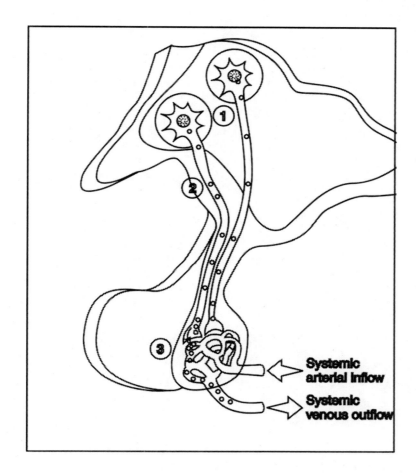

311

18-7 Vascular link between the hypothalamus and anterior pituitary.

Color

Hypothalamus

Neurosecretory neurons in hypothalamus

Systemic arterial blood in hypothalamichypophyseal portal system

Posterior pituitary

Anterior pituitary

Systemic venous blood out

Capillaries in anterior pituitary

Capillaries in hypothalamus

Endocrine cells of anterior pituitary

Releasing and inhibiting hormones

Hypophysiotropic hormones

Anterior pituitary hormone

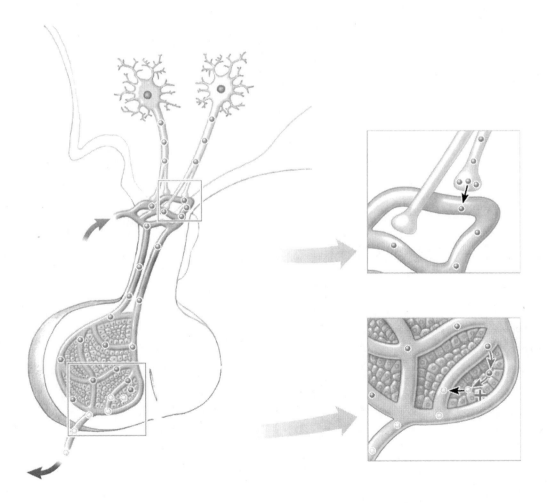

18-9 Anatomy and growth of long bones.

(a) Anatomy of long bones. (b) Two sections of the same epiphyseal plate at different times, depicting the lengthening of long bones.

Color

Articular cartilage
Bone of epiphysis
Resting chondrocytes
Epiphyseal plate
Bone of diaphysis
Marrow cavity

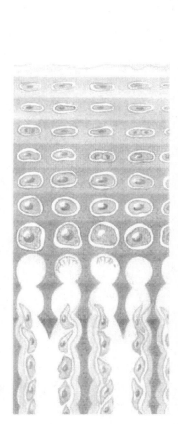

Label

Chondrocytes
    undergoing cell division
Older chondrocytes
    enlarging Calcification of
extracellular matrix (entrapped
    chondrocytes die)
Dead chondrocytes
    cleared away by
    osteoclasts
Osteoblasts
    swarming up
    from diaphysis and
    depositing bone
    over persisting
    remnants of
    disintegrating
    cartilage

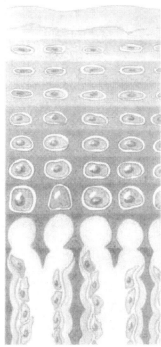

315

19-2 Synthesis, storage, and secretion of thyroid hormone.

Color

Blood

Thyroid follicular cell

Lysosome

Endoplasmic reticulum

Golgi complex

Colloid

List the 9 steps in this process

1. _____
2. _____
3a. _____
3b. _____
4a. _____
4b. _____

5. _____
6. _____
7a. _____
7b. _____
8. _____
9a. _____
9b. _____

Insert all the chemicals into the diagram below.

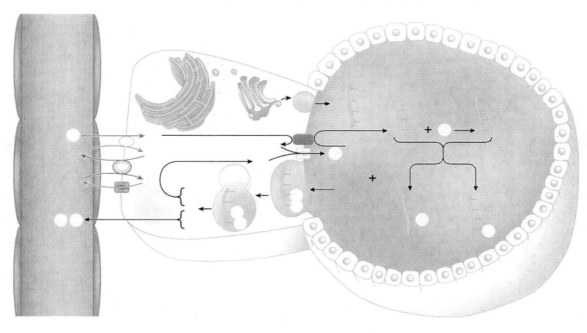

19-7 Anatomy of the adrenal glands.

(a) Location and structure of the adrenal glands. (b) Layers of the adrenal cortex.

Color

Connective tissue capsule

Zona glomerulosa

Medulla

Zona reticularis

Cortex

Zona Fasciculate

Label

Cortex

Glucocorticoids and
    Sex hormones

Mineralocorticoids

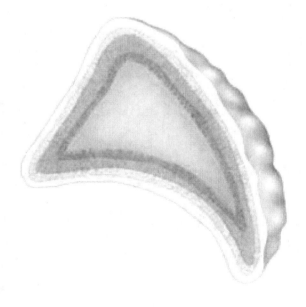

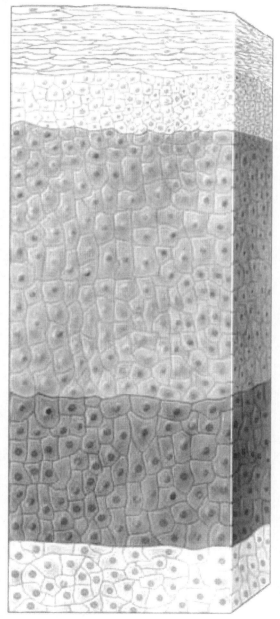

19-15 Location and structure of the pancreas and cell types in the islets of Langerhans.

Color

Duct cells secrete aqueous NaHCO$_3$ solution

Acinar cells secrete digestive enzymes

Exocrine portion of pancreas (acinar and duct cells)

Endocrine portion of pancreas (islets of Langerhans)

Hormones (insulin, glucagon)

Blood

B cell $\Rightarrow$ insulin

Exocrine cells

Islet of Langerhans

α cell $\Rightarrow$ glucagon

D cell $\Rightarrow$ somatostatin

Capillary

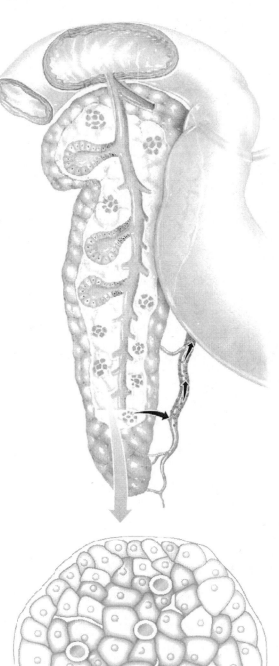

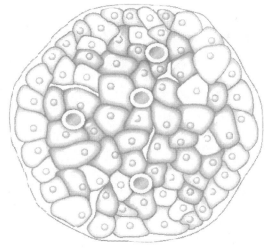

19-17 Stimulation of insulin secretion by glucose via excitation-secretion coupling.

Color

ATP-sensitive K⁺ channel
Voltage-gated Ca²⁺ channel
Insulin vesicle
Insulin molecule ATP
GLUT-2
β cell

Label

Glucose
Glucose-6-phosphate
ATP
Ca2+
Potassium in cell because less leaves

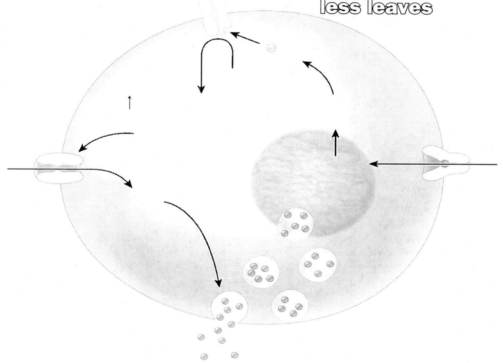

List the 9 steps in this process

1. _____
2. _____
3. _____
4. _____

5. _____
6. _____
7. _____
8. _____
9. _____

19-23 Organization of compact bone into osteons

(a) Structure of a long bone showing location of compact bone and trabecular bone. (b) An osteon, the structural unit of compact bone, consists of concentric lamellae (layers of osteocytes entombed by the bone they might have deposited around themselves) surrounding a central canal containing a small blood vessel branch. The light micrograph is of a compact bone in a human femur (thigh bone). (c) A magnification of lamellae.

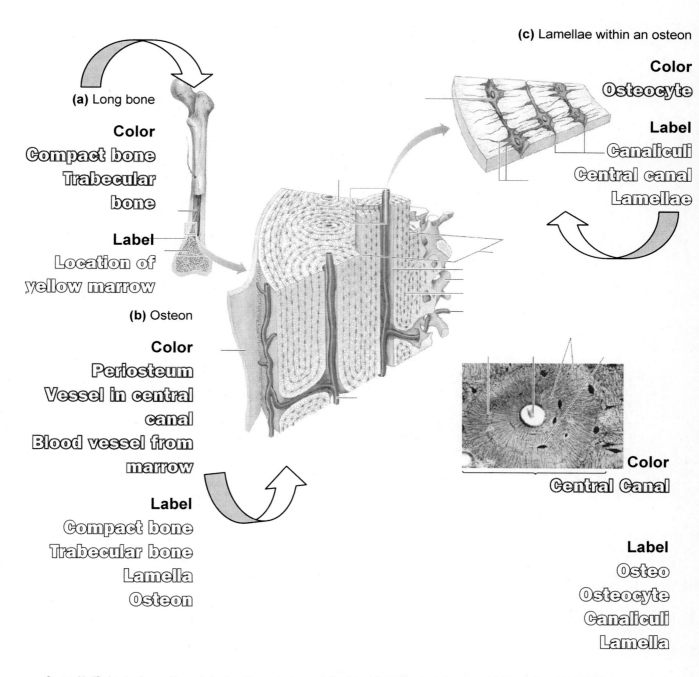

(c) Lamellae within an osteon

Color
Osteocyte

Label
Canaliculi
Central canal
Lamellae

(a) Long bone

Color
Compact bone
Trabecular bone

Label
Location of yellow marrow

(b) Osteon

Color
Periosteum
Vessel in central canal
Blood vessel from marrow

Label
Compact bone
Trabecular bone
Lamella
Osteon

Color
Central Canal

Label
Osteo
Osteocyte
Canaliculi
Lamella

Source: Modified and redrawn with permission from Human Anatomy and Physiology, 3rd Edition, by A. Spence and E. Mason. Copyright © 1987 by the Benjamin/Cummings Publishing Company. Reprinted by permission of Pearson Education, Inc.

19-24 Fast and slow exchanges of Ca2+ across the osteocytic-osteoblastic bone membrane.

(a) Schematic representation of the osteocytic- osteoblastic bone membrane. The entombed osteocytes and surface osteoblasts are interconnected by long cytoplasmic processes that extend from these cells and connect to one another within the canaliculi. This interconnecting cell network, the osteocytic- osteoblastic bone membrane, separates the mineralized bone from the plasma in the central canal. Bone fluid lies between the membrane and the mineralized bone. (b) Schematic representation of fast and slow exchange of Ca2+ between the bone and plasma.

Color

Mineralized bone

Osteocyte

Osteocytic-osteoblastic
bone membrane

Canaliculi

Bone fluid

Oesteoblast

Central canal

Outer surface

Plasma

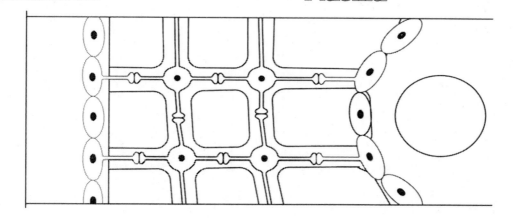

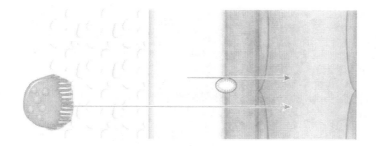

20-1 The male reproductive system.

(a) The pelvis in sagittal section. (b) Posterior view of the reproductive organs. Portions of some organs have been removed.

Color

Anus

Bulbourethral gland

Cords of erectile tissue

Ductus deferens

Ejaculatory duct

Epididymis

Glans penis

Penis

Prostate gland

Public bone

Rectum

Scrotum

Seminal vesicle

Testis

Ureter

Urethra

Urinary bladder

Vertebral column

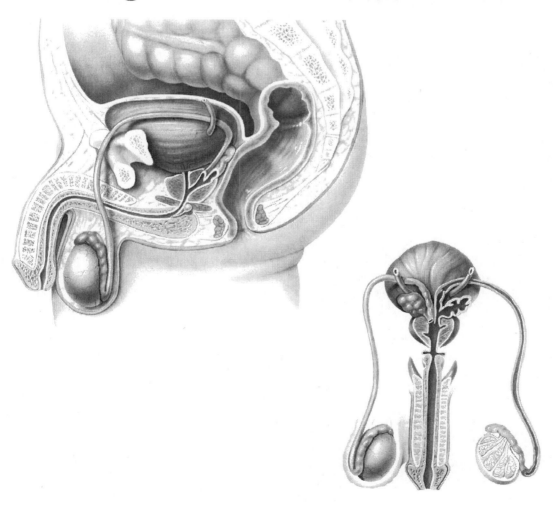

20-2 The female reproductive system.

(a) The pelvis in sagittal section.

Color

Oviduct

Ovary

Fimbriae

Uterus

Urinary bladder

Pubic bone

Urethra

Clitoris

Labium minora

Labium majora

Vagina

Rectum

Cervix

Anus

Vertebral column

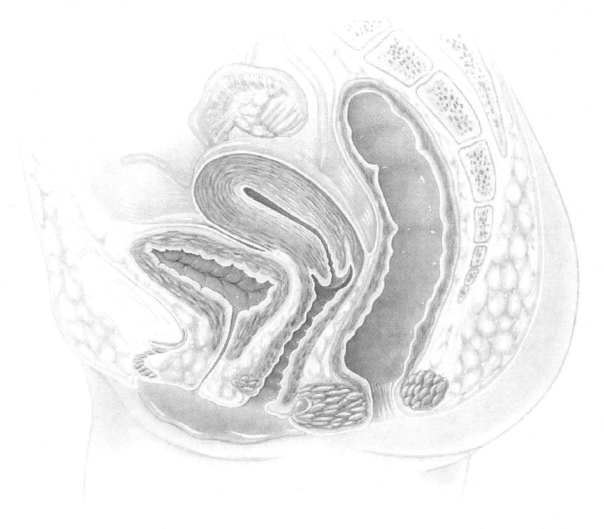

20-2 The female reproductive system. (cont.)

(b) Posterior view of the reproductive organs. (c) Perineal view of the external genitalia.

Color

Labium minora

Labium majora

Oviduct

Endometrium

Myometrium

Vagina

Cervical canal

Cervix

Uterus

Ovary

Ovarian vessels

Fimbriae

Clitoris

Opening of urethra

Hymen

Vaginal opening

Perineu

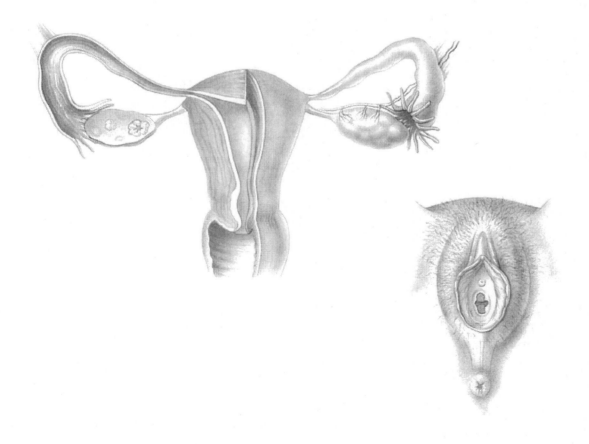

20-4 Sexual differentiation of the external genitalia.

Color

Anal opening

Anus

Clitoris

Developing clitoris

Developing penis

Genital (labial) swelling

Genital (scrotal) swelling

Genital swellings

Genital tubercle

Labia majora

Labia minora

Near term

Urethral fold

Urethral fold (partially fused)

Urethral opening

Vagina

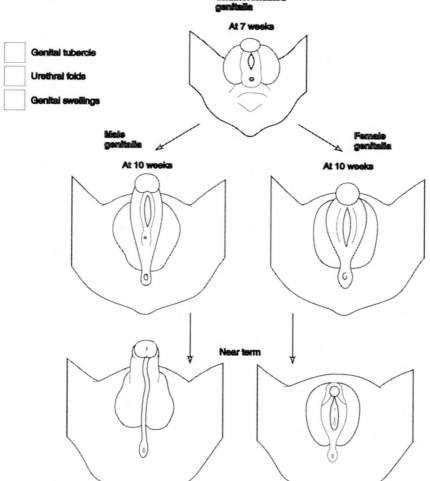

20-5 Sexual differentiation of the reproductive tract.

Color

Müllerian ducts

Undifferentiated gonads

Wolffian ducts

Ovaries

Oviducts (Fallopian tubes)

Uterus

Vagina

Fimbria Epididymis

Testes

Ductus deferens

Seminal vesicles

Prostate

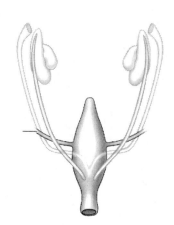

Label

Undifferentiated reproductive system

Wolffian ducts degenerate

Müllerian ducts degenerate

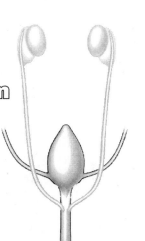
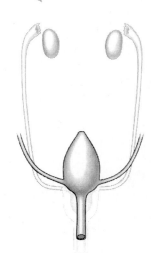

20-6d Testicular anatomy depicting the site of spermatogenesis.

(d) Relationship of the Sertoli cells to the developing sperm cells.

Color

Lumen of seminiferous tubule

Spermatozoon

Sertoli cell

Spermatids

Secondary spermatocyte

Primary spermatocyte

Tight junction

Spermatogonium

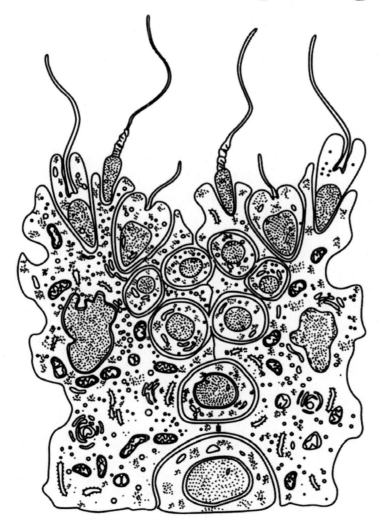

20-7 Spermatogenesis.

Color

Spermatogonia
Spermatogonium
Primary spermatocytes
Secondary spermatocytes
Spermatids
Spermatozoa
Mitotic proliferation
Meiosis
Packaging

Label

First meiotic division
Second meiotic division
Mitotic proliferation
Meiosis
Packaging
The number of chromosomes in each stage of spermatogenesis

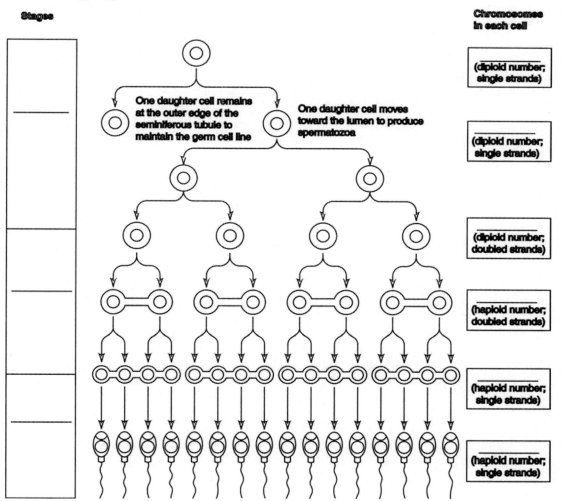

Stages

Chromosomes in each cell

One daughter cell remains at the outer edge of the seminiferous tubule to maintain the germ cell line

One daughter cell moves toward the lumen to produce spermatozoa

(diploid number; single strands)

(diploid number; single strands)

(diploid number; doubled strands)

(haploid number; doubled strands)

(haploid number; single strands)

(haploid number; single strands)

20-8b Anatomy of a spermatozoon.

(a) A phase-contrast photomicrograph of
human spermatozoa. (b) A spermatozoon has three functional parts: a head with its acrosome
"cap," a midpiece, and a tail.

Color

Acrosome

Nucleus

Mitochondria

Microtubules

Label

Midpiece

Tail (flagellum)

Head

343

Figure 20-9 Control of testicular function.

Color

Label

Kisspeptin
GnRH
LH
FSH
Inhibin
Spermatogenesis
Testosterone
Masculinizing effects

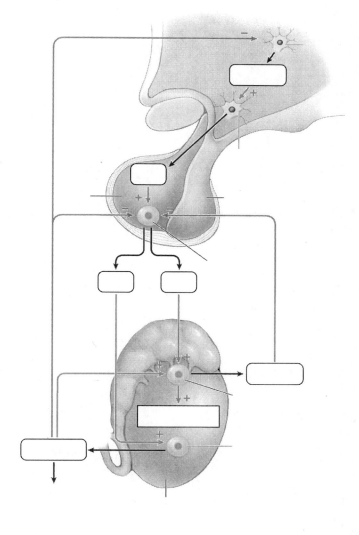

20-12 Oogenesis.

Color

Oogonium
Primary oocytes
Enlarged primary
oocyte
Secondary oocyte
Mature ovum

Label

Mitotic proliferation prior
  to birth
Meiosis
Number of chromosomes
  in each cell
Second polar body
Polar bodies degenerate
First polar body

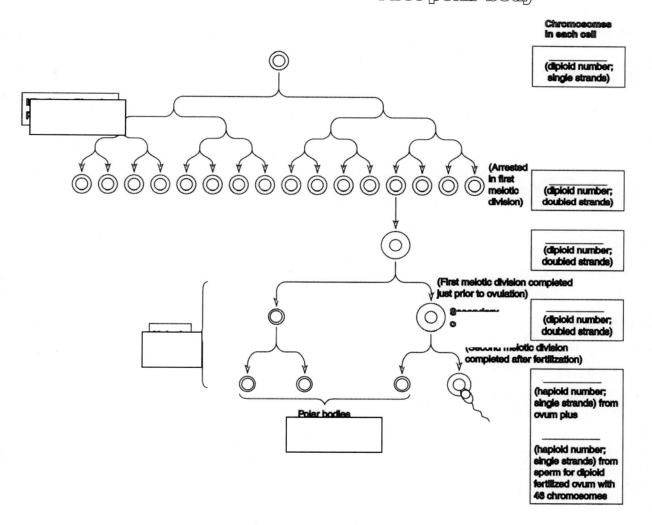

Chromosomes
in each cell

(diploid number;
single strands)

(Arrested
in first
meiotic
division)

(diploid number;
doubled strands)

(diploid number;
doubled strands)

(First meiotic division completed
just prior to ovulation)

(diploid number;
doubled strands)

(Second meiotic division
completed after fertilization)

(haploid number;
single strands) from
ovum plus

(haploid number;
single strands) from
sperm for diploid
fertilized ovum with
46 chromosomes

Polar bodies

20-14 Ovarian Cycle.

(a) Ovary showing progressive stages in one ovarian cycle. All of these stages occur sequentially at one site, but the stages are represented in a loop in the periphery of the ovary so that all of the stages can be seen in progression simultaneously.

Color

Antrum

Corona radiate

Corpus luteum

Degenerating corpus luteum

Developing corpus luteum

Follicular cells

Follicular remnant

Granulosa cells

Ovarian surface

Ovulated ovum (secondary oocyte)

Ovum (primary oocyte)

Ovum (secondary oocyte)

Primary follicle

Primary oocyte

Secondary follicle

Thecal cells

Zona pellucida

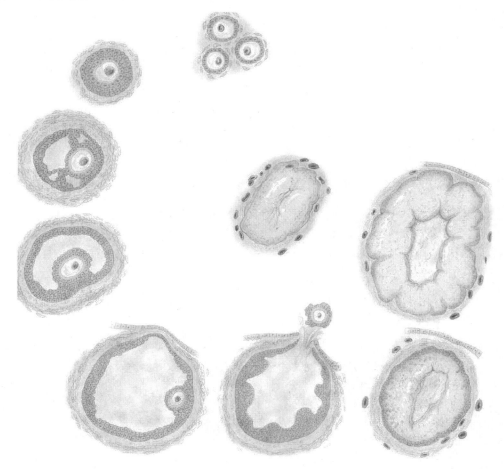

20-15 Comparison of mitotic and meiotic divisions producing spermatozoa and eggs from germ cells.

Color

Male germ cell
Permatogonium
Primary spermatocyte
Primary oocyte
Secondary oocyte
First polar body
Ootid
Mature ovum
Secondary spermatocyte
Spermatid
Sperm
Female germ cell
Oogonium

Label

Meiosis I
Meiosis II
Mitosis
From first
    polar body
Second polar
    body
Polar
bodies
disintegrate

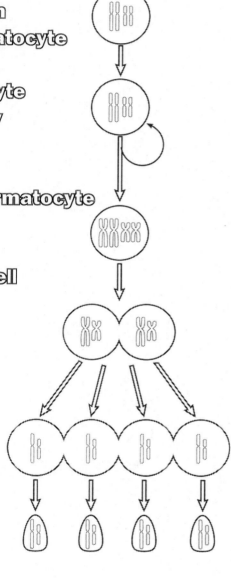

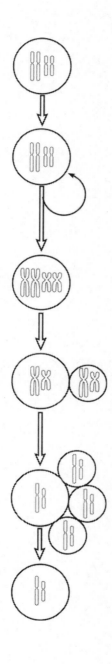

Figure 20-20 Control of the LH surge at ovulation.

Color

Kiss1 neuron in anteroventral periventricular (AVPV) nucleus

Hypothalamus

GnRH-secreting cell

Anterior pituitary

Posterior pituitary

Gonadotrope

Mature follicle

Ovary

Label

Kisspeptin	FSH	Inhibin	LH
GnRH	Ovulation	High levels of estrogen	

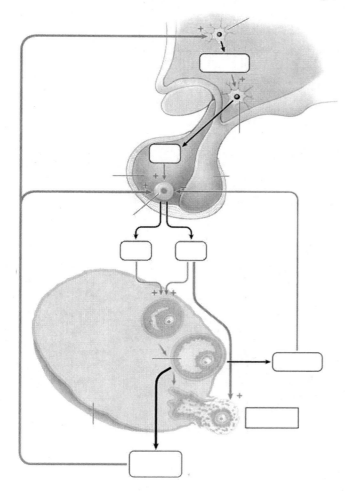

Figure 20-21 Feedback control during the luteal phase.

Color

Kiss1 neuron in AVPV nucleus

Kiss1 neuron in ARC nucleus

Hypothalamus

GnRH-secreting cell

Anterior pituitary

Posterior pituitary

Gonadotrope

Ovary

Corpus luteum

Label

Kisspeptin

High levels of estrogen

GnRH

High levels of progesterone

LH

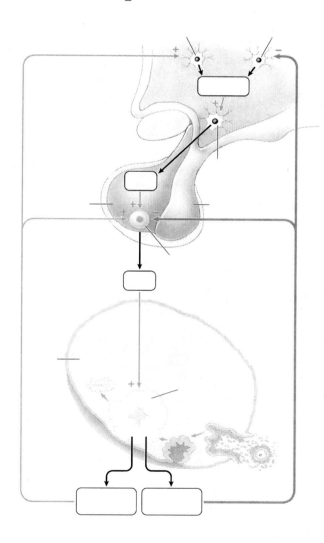

20-22 Ovum and sperm transport to the site of fertilization.

Color

Oviduct

Fimbria

Ovary

Uterus

Cervical canal

Vagina

Penis

Ovulated ovum

Label

Ampulla of oviduct

Optimal site of fertilization

Sperm surrounding ovum

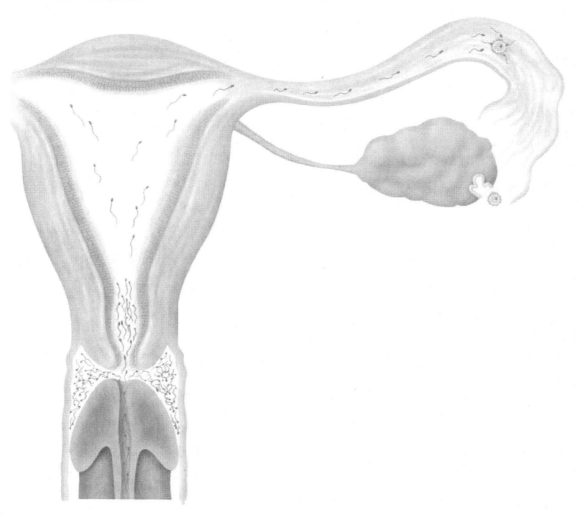

20-23a Process of fertilization.

(a) Sperm tunneling the barriers surrounding an ovum.

Color

Zona pellucida

ZP3 binding site

Corona radiate

Ovum cytoplasm

Sperm midpiece and tail

Sperm nucleus

Ovum plasma membrane

Sperm plasma membrane

Cytoplasm of ovum

First polar body

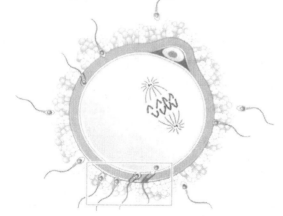

Label

Nucleus of ovum
   undergoing
   second meiotic division

Spermatozoa

Acrosomal vesicle

Cortical granules

List the 5 steps of this process

1. _____
_____

2. _____
_____

3. _____
_____

4. _____
_____

5. _____
_____

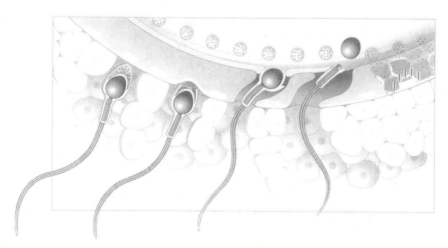

20-24 Early stages of development from fertilization to implantation.

Note that the fertilized ovum progressively divides and differentiates into a blastocyst as it moves from the site of fertilization in the upper oviduct to the site of implantation in the uterus.

Color

Morula

Endometrium of uterus

Cleavage

Spermatozoa

Fertilization

Ovum

Ovary

Secondary oocyte

Trophoblast

Inner cell mass

Label

Polar bodies

Ovulation

Implantation

Blastocyst

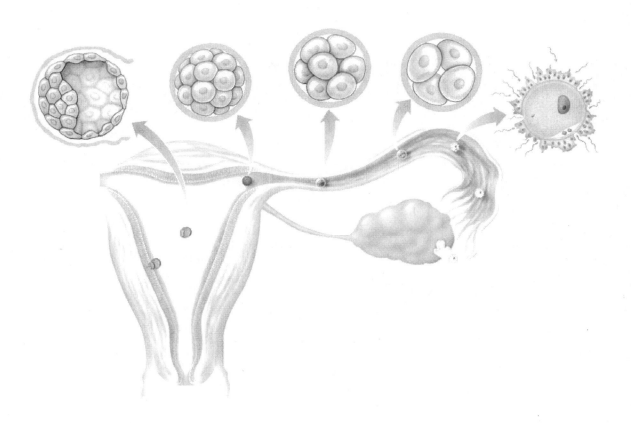

20-25 Implantation of the blastocyst.

Color
Endometrium
Capillary
Inner cell mass
Surface of uterine lining
Cords of trophoblastic cells
Trophoblast
Uterine cavirty
Decidua
Start of amniotic cavity
Developing embryo

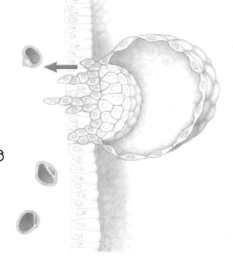

List the 3 steps of this process

1. _____
   _____
   _____
   _____

2. _____
   _____
   _____
   _____

3. _____
   _____
   _____
   _____

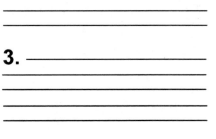

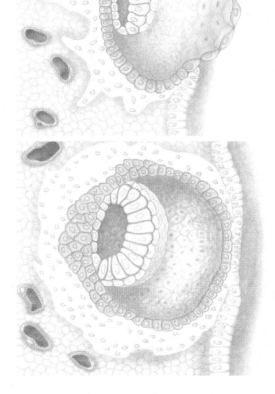

20-26a Placentation.

Fingerlike projections of chorionic (fetal) tissue form the placental villi, which protrude into a pool of maternal blood. Decidual (maternal) capillary walls are broken down by the expanding chorion so that maternal blood oozes through the spaces between the placental villi. Fetal placental capillaries branch off the umbilical arteries and project into the placental villi. Fetal blood fl owing through these vessels is separated from the maternal blood by only the capillary wall and thin chorionic layer that forms the placental villi. Maternal blood enters through the maternal arterioles, then percolates through the pool of blood in the intervillus spaces. Here, exchanges are made between the fetal and maternal blood before the fetal blood leaves through the umbilical vein and maternal blood exits through the maternal venules.

Color

Maternal arteriole

Maternal venule

Amniotic fluid

Chorion/amnion

Uterine decidual tissue

Umbilical vein

Umbilical arteries

Placental villus

Label

4 weeks

8 weeks

12 weeks

Full term

Placenta

Umbilical cord

Chorionic tissue

Intervillus space

Fetal vessels

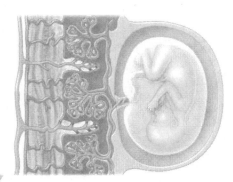

20-31 Mammary gland anatomy.

The alveolar epithelial cells secrete milk into the lumen. Contraction of the surrounding myoepithelial cells ejects the secreted milk out through the duct. (a) Internal structure of mammary gland, lateral view. (b) Alveolus within mammary gland

Color

Adipose tissue

Lumen

Duct

Myoepithelial cell

Alveolar epithelial cell

Lobule containing alveoli

Label

Nipple

Duct

Milk

Secretion

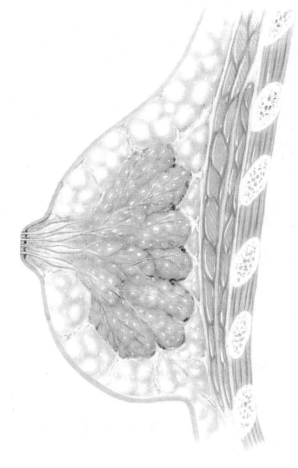

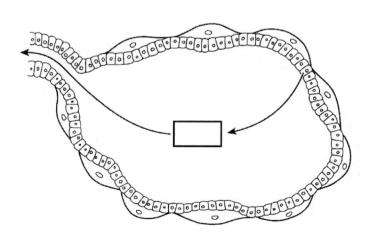

CPSIA information can be obtained
at www.ICGtesting.com
Printed in the USA
FFOW03n0516180215
11205FF